THE CARVING OF CANADA

Sculptor Eleanor Milne, the 'Ti'Elen of the tale

The Carving *of* Canada

A TALE OF PARLIAMENTARY GOTHIC

To Julie Mason,
NDP Chief of Staff
who makes her own
creative contribution to
"The Carving of Canada".

Munroe Scott

Munroe Scott
Nov. 23/99

PENUMBRA PRESS

© Copyright Munroe Scott and Penumbra Press, 1999

All rights reserved. No part of this book covered by the copyrights hereon may be reproduced or used in any form or by any means — graphic, electronic, or mechanical — without the prior written permission of the publisher. Any request for photocopying, recording, taping, or storage or retrieval systems of any part of this book shall be directed in writing to CanCopy, Toronto.

CANADIAN CATALOGUING IN PUBLICATION DATA

Scott, Munroe, 1927-
    The carving of Canada : a tale of parliamentary gothic

ISBN 0-921254-93-8

    1. Milne, Eleanor   2. Parliament buildings (Ottawa, Ont.)— History.
I. Scott, Ian D.   II. Title
NB249.M56S36   1999   730'.92   C99-901350-5

In loving memory of Hilda Mary, who is with me "in the deep heart's core."

Mythology is the symbolism of common tribal experience. It is the interior road map of Canada, the music of Canada we dance to even when we cannot name the tune. Without a mythology – "the maps lost, the voyages cancelled" – there is no country.

— Michael Valpy, *Globe & Mail*, 9 July 1990

CONTENTS

  8 Acknowledgements
  9 Preface

 11 Destruction and Rebirth
 25 Creation Phase 1
      47 Phase 2
      57 Phase 3
      65 Phase 4
      81 Phase 5
 89 Celebration Party 1
    101 Party 2
    115 Denigration
 125 Unification Step 1
    139 Step 2
    165 Evolution

175 Index of Carvings and Windows

ACKNOWLEDGEMENTS

Photographs of the stain glass windows on pages 111-114 are reproduced with the consent of the Library of Parliament / reproduit avec permission de la Bibliothèque du Parlement. Photographs of Eleanor Milne on the frontispiece and page 24 are copyrighted and provided by Dominion Wide Photograph Limited and Bill Olson, Ottawa. The photograph on page 15, "Full scale construction of Centre Block showing Library, Ottawa. 15 August 1916," is from the National Archives of Canada Department of Public Works Collection, PA130624. Author photo on back cover is by Vail. All other photographs of carvings are Copyright © Ian D. Scott.

PREFACE

This book began several years ago as a biography of Eleanor Milne, who, at the time, was Canada's Parliamentary Sculptor. It transformed itself into "a tale of Parliamentary Gothic," a story for all Canadians of any age who still enjoy a good yarn. If it encourages my countrymen to pay more attention to the symbolism of Parliament Hill and to the role of the artist in national life, then I am happy. If it encourages someone more skilled than myself to write a more orthodox biography of Ms Milne then I am more than happy.

I would like to thank sculptor Eleanor Milne, glassmaster Russell Goodman, carvers Maurice Joanisse, Fernand Rossignol, Christopher Fairbrother, architect Robert Calvert, and others too numerous to mention who allowed themselves to be interviewed. Initially, the Speaker of the House of Commons, the Hon. John Fraser, gave me full access to the masterworks, and later his successor, the Hon. Gilbert Parent, and his staff were enormously cooperative in giving access to myself and to my photographer. The colour photos of the windows were graciously made available by Mr. Mike Graham of the Parliamentary Library. Some eight years ago the Ontario Arts Council contributed modest seed money, thereby boosting my morale, for which I thank it. I give special thanks to freelancer Ken Dessin and to Audrey Dubé, Document Curator of the House of Commons, for extensive assistance with research documentation, and I would be greatly remiss not to express my gratitude to my editor, Douglas Campbell, for his insightful guidance.

Contrary to custom I do not apologize for "errors or omissions." This is, after all, a tale told beside a campfire.

— M.S., Peterborough, 1999

DESTRUCTION AND REBIRTH

Once upon a future, years from now, a small fire sparkled on the granite shore of a northern lake.

Girl sat by the fire and looked across the flames toward rainbow waters and silhouette islands. It had been almost an hour since Sun Artist had washed brushes and gone to rest but the sky easel was still stained with colours. In the mirror of the lake Girl could see that Moonship was already setting sail for the night. She was thinking of nothing, and embracing everything. Her spirit was abroad, bathing in infinite beauty.

Youth was lying nearby, head resting on a small log, eyes fixed on the evening sky. He had watched the Venus Star emerge on the western horizon and had finally spotted Mars. He was now awaiting the arrival of the Great Bear and hoping to spot the Hunter, the Bull, and the Lion. He was wondering how his body could be so small and his mind so immense.

By the fire itself, Boy was busily pushing marshmallows onto the sharpened end of a stick, placing them over the flames for a swift toasting, and eating them before they had a chance to swell and gleam brown. He changed the single stick for one with forked prongs that could hold several morsels at once. The technical improvement seemed to please him and he ate on.

Girl glanced toward the tent as Old One lifted the flap and emerged. "Good," she thought, "Old One has had an after-meal nap. If dreams have come, stories will follow."

"It seems to me," said Youth, speaking to no one in particular, "that if the universe is exploding outward, then there must, somewhere, be a centre to the universe."

Old One moved in on quiet feet and settled easily onto a log. "Ah yes, but perhaps the universe is an Idea. Then the centre of the universe would be a Mind."

"Do you believe that?" asked Youth.

"One can wonder without believing. Belief requires either proof or faith – but History is full of people who believed without either."

Boy stuffed two marshmallows in his mouth and mumbled.

"I heard," said Old One, "but didn't understand."

Boy swallowed quickly. "I said, who made History?"

"That," said Girl, "is like asking what came before History."

Youth laughed, "Or was the North Pole lost before it was found?"

"How far south does North come?" asked Girl.

"By travelling west you can arrive in the East," said Youth. "Where does one become the other?"

"Wait, wait, wait!" laughed Old One, "do you expect answers?"

"Yes," said Youth.

"Tell us a story," said Boy.

"Now you're talking," said Old One.

"But is it a true story?" Girl preferred fantastical tales that were true.

Old One hesitated a moment, memory wheeling in circles like a hunting hawk. "It is carved in stone. It is mounted in glass."

Boy frowned, Youth shrugged, Girl smiled, and Old One began.

*****

The Firebird came with coals in its talons, swooping through the early mists of time, and in the heavens kindled a great fire fuelled by the fibres of its own nest, and out of the flames the world was born and on the world all manner of life was born and out of life love was born. But fire was the beginning.

It is no accident that the tale you are about to hear, this story of the marvellous creation of a strange and exotic country by a girl grown to woman, this chronicle of heartbreak and healing, it is no accident that it, too, begins with a fire.

Much of Europe was on fire in those days. There, most of the mighty nations of the world were at each other's throats, aided by their friends and neighbours and overseas dominions. Warriors had been marching to and fro, guns flaming and cannons roaring, and whole armies had dug themselves into the ground like angry badgers and they glared at each other across smoking wastelands.

But there was no war on Ottawa's parliamentary Hill. The rulers on The Hill had sent soldiers to the war because devastation is something one prefers to deliver rather than to receive, but the war was oceans away and on The Hill it was a peaceful winter's evening. Even so, all it took was a small spark on a bureaucratic paper in a legislative wastebasket and fire came to The Hill. The flames licked from the basket, ignited the lacquer of a desk, flared upward on drapes and surged into the hallways.

In the elegant Red Chamber the Ancient Ones roused themselves and fled. In the Green Chamber a People's Orator paused in the midst of an exhortation to the populace to eat more fish, and watched as his fellows also fled. Scribes, guards, pages, bureaucrats, factotums, mandarins, functionaries and supplicants all fled. Firemen and policemen came to help but pine-panelled walls roared into vapour, glass melted, the throne turned into charcoal, corridors collapsed, ornate doors disintegrated, walls of limestone split, and the flames climbed high into the clock tower as a big bell began the count of midnight. On the stroke of twelve the great bell fell, down, down, down, and the building that had symbolized the hopes of many people became ashes and rubble, a desolation of despair, while the great river that passed behind The Hill continued to flow between solitude and solitude.

The next day, as the cold sun rose on a frozen February morning, as the smoke lifted and the steam evaporated, a strange sight greeted wondering eyes. Either something had not burned or something new had arrived. Behind the rubble, close to the cliff edge that shaped the back of The Hill, stood a large, round, towering, cone-shaped object. There were those who said it had been there all along, while others said it had never looked like this.

In those far-off days that round structure, buttressed with arched supports flying down from roof to ground, looked like a Gothic survival from a distant

DESTRUCTION AND REBIRTH

past. But you, a later generation, seeing it isolated and smoking, might have thought of a nose cone from a distant future.

The outer skin of the Cone glowed with heat and wisps of steam still rose from its blackened sides. Was the heat from the flames of the fire? Or was it residual heat from an atmospheric entry? You and I, looking back, are permitted to wonder.

Mandarins who ventured into the apparently hollow interior of the structure came out babbling to themselves and the only words that made sense were "memory" and "dreams." But is memory not of the past? Are dreams not of the future? Could the strange alchemy of heat have warped both past and future into the present?

So amazing was this edifice, so astounding in its solitary simplicity, so soaring in its aspirations, so solid on its foundations, that, to prevent the common people from entering, it was immediately secured and sealed. A guard, wearing a red uniform and a heavy winter coat and a carrying a heavier firearm, was placed on duty at the door of the structure the flames had so miraculously isolated and revealed. Memory and dreams were under lock and key, inaccessible to all but the privileged.

The chronicles tell that the faraway war between republics and empires and principalities raged on for two more years and that by the time an armistice was declared the warriors sent from The Hill had, through bravery, dedication, and sacrifice, forged a sense of family. Thus it was that war, the most imbecilic and destructive of mankind's activities, created among the sons and daughters of The Hill not only a sense of belonging but even of purpose. And just as a sense of identity grew out of the ashes of a great war so did a new building grow out of the rubble on The Hill.

More than half a century before, the first building had been hailed as a builder's triumph. There was, so it was said, "no modern Gothic purer of its kind," although it was not clear how anything could be modern and Medieval and, at the same time, "pure."

But that building, pure or not, had now burned.
There were those who had thought the original building to be far too big. "Such monstrous folly," they had said, "was never perpetrated in this world before."
But that building, folly or not, was now rubble.
A new structure rose in its place that was even more splendid. Taller than the first one, more commodious, more graceful, it was a building that cried out to embrace a nation. What had been said with exaggeration of the first building could now be said with truthfulness of this one, that it was "characterized by purity of art, manliness of conception, beauty of outline, and truthful nobility of detail."
Ah, the detail.
There is more to a fine building than mere materials, although when the builders and the masons and carpenters and the tile layers and the plumbers and the glaziers and the roofers and the ironmongers and the steel workers and all the other trades people had departed, the materials were indeed marvellous. There was solid grey native limestone, and western limestone bearing fossils from ancient seabeds. There was soaring sandstone from local quarries. There was marble from Belgium and stone from France. There was bronze and linen and gold, there was ebony and walnut and oak. And even as the building grew the carvers had been at work.
In the elegantly proportioned Green Chamber, where the Orators debated, the balcony parapets, made of fine wood, were embellished at the base with borders of leafy vines entwining animals and birds, the offspring of a bountiful nature. The posts were capped

with figures of animals, children, and gnome-like people. On the building's outer walls, stone-fleshed goats and bears and creatures more strange crouched on doorway gables and lurked in elevated crannies. Great stone gargoyles peered down from a three-hundred-foot tower and on lesser towers and walls grotesque stone bodies carried recognizable caricatures of local architects and foreign politicians.

For years after the builders had gone mysterious men came both by day and by night and the inner foyers and hallways rang to the sound of mallet and chisel. Strange men wearing smocks and dust-covered overalls climbed ladders and scaffolds and hewed skilfully at archways, pillars, capitols, corbels and blocks. The building began to grow in spirit. The sages and the seers watched wonderingly, saying among themselves that there were symbols and seeds here that might some day blossom into more than a mere building.

There were philistines on The Hill in those days, men lacking in sentiment, culture, and vision, who authorized wages for carvers and then interfered in their work. A sculptor created a memorial in white marble to honour nurses who had offered up not only their compassion but often their lives. The central figure was of a woman in flowing robes, head bowed in sympathy, arms outstretched to shelter the sick and the wounded. The King's First Minister, perhaps seeing in the face a trace of his mother, ordered the nose to be suitably altered.

Inside the Great Tower, which itself was dedicated to Peace, was a chamber in memory of the sons and daughters of The Hill who had died in faraway wars.

A carver, inspired by their deeds, fell to with a will. He ignored the Gothic foliage beloved by the Chief Architect, foliage that was already beginning to engulf the building, and instead carved epic events. Now established in the annals of The Hill were battalions moving into line in the fields of Flanders, smoking howitzers at Arras, field guns rolling up to support attacking infantrymen. Tapping chisels carved foreign names to emblazon them forever in the chronicles of a new land – strange names like Ypres, Passchendaele, Beaumont Hamel, and Vimy, names that had been ribboned together with blood.

The philistines, who viewed art as an ornament, and craft as a commodity, inspected the bottom line of their ledger books and begrudged craftsmen their wages. The strong-minded carver who had entrenched the saga of a worldwide war into the chronicle of The Hill was denied his back pay. Soon, other carvers were dismissed. Before long the mallets and steel sang no more.

Behind the building the strange cone-like structure, now almost hidden, continued to store entrapped memories and suppressed dreams. At its doorway, now linked by a noble hallway to the main building, there still stood an armed guard, keeping the people from intruding into the sanctuary of their own memories.

Years passed and there was drought, desolation and despair throughout the vast territories belonging to The Hill. In the town around the base of The Hill the police guarded the garbage midden to keep the people from living there. In other towns millionaires dined while poor people starved. Men were shot while looking for work. In the Green Chamber a People's Orator rose to his feet and verbally attacked the philistines who judged the world through the dollar signs on the bottom lines of their ledgers. "Strict economy on the part of this government," he said, "will mean less work and more unemployment, more hunger, more destitution, more want, more disease, more crime, more insanity, more suicides and more human deterioration."

One of the most skilled of carvers, now ageing, weary, and long unemployed, came cap in hand to The Hill looking for work and, miracle of miracles, he received it. Dust storms still raged, unemployed workers still roamed the solitudes, bread lines still stretched from soup kitchen to soup kitchen, but master carver Cleophas Soucy picked up mallet and chisel and once again the stones on The Hill began to ring.

Cleophas carved a most intricate, delicate, and elegant yet substantial design into the great archway framing the doors that led to the Red Chamber where the Ancient Ones gathered to meditate. Carving with boldness and skill, he created a rough texture on even the most delicate of tracery so there would be both shadow and substance to impress itself upon the eye.

The music of the mallet sang out in the building's rotunda as Cleophas surrounded the groined arches of limestone with symbols of a living land populated by owls, Eskimos, hawks, Indians, beavers, trappers, bears, woodsmen, chickadees, sailors, and wolves, all living in harmony.

A regal lion and a noble unicorn appeared in the open air, flanking the archway at the base of the great tower, symbols of history and mythology guarding a single entrance. They were the work of another carver, known by the name Coeur de Leon MacCarthy. Coeur de Leon moved inside, passed through the elegant doorway of Cleophas and on into the austere Red Chamber. There, in a niche over the throne, he carved a marble bust of the long dead Queen who had bestowed both power and privilege upon The Hill.

Cleophas carried his tools into the open air and planned to carve a symbolic family of ten beavers atop the central archway that led past Coeur de Leon's lion and unicorn and into the tower. The philistines declared that beavers did not have large families and issued orders for Cleophas to be content with one beaver. The populace howled protest and hurled letters and editorials at the bureaucrats. Cleophas, seeing orders and missives flying in all directions, carved one amazed beaver sheltering behind a large shield.

The years went by and eventually Cleophas and Coeur de Leon passed away and a new Master Carver worked in their place. He did not like the texture of the work of Cleophas, the texture that sang to the eye. While not daring to touch the Red Chamber doorway, he had his men chisel a smoother surface upon much of Cleophas' work. He carved in less depth and had his men paint the grooves a dark grey, giving an illusion of deep shadows and of having done what was not done. Violating modesty he carved his own face upon a block where it could peer down onto kings, queens, and first ministers. The philistines were not all among the bureaucrats.

In the course of all the long years that had passed since the fire a king had died, a king had resigned, a new king had reigned, first ministers had come and gone, a second worldwide war had brought desolation to more faraway places, again a king had died, and now a queen reigned as Time and the river continued to flow past The Hill. But still the building was not finished. It only appeared to be finished.

In the foyers, corridors and chambers most of those bewildering devices that combine to compose the poem of Gothic architecture, the bosses, spandrels, fenials, corbels, and groins, were embellished with vines intertwined with Tudor Roses and Fleurs de Lys, with Scotch Thistles and Irish Shamrocks, with maple leaves and pine cones and shields, with beavers, barristers, hunters, trappers, tradesmen, traders, explorers, and even journalists. But something was missing.

Astute Bureaucrats posted a notice to the effect that the Master Carver was retiring and was to be replaced by a fully qualified sculptor.

Many applied.

One was chosen. A woman. Her name was Eleanor.

She was in the early prime of life, light of complexion, fair of hair, blue of eye, slim of figure, with an easy manner and a quick smile. On The Hill she was to take charge of a domain that had been all male.

The carvers had always been men. Mostly short, powerfully built, stocky men. They were now to be led not by a carver but by a sculptor. And the sculptor was a woman. And the woman was six feet tall.

The eldest of the carvers, father to three of them, looked at the tall young artist and dubbed her, with a French twinkle, "'Ti'Elen" – "Little Elen."

The Master Carver, who had not been consulted in the selection of his successor, took 'Ti'Elen around the corridors of The Hill, showed her the accomplishment of the years, introduced her to her men, showed her the carver's workshop deep in a sub-basement in the bowels of the building below the Red Chamber, and then left. 'Ti'Elen was on her own.

Or was she?

In another part of the building, at the foot of the Hall of Honour that led from the principal rotunda to the ornately carved doors of pine that gave access to the mysterious pyramidal Cone at the rear, something was missing. At some time during the years, exactly when the chronicles do not say, the armed guard with the red tunic had been quietly removed. Not only was it easier to enter the wonderful area beyond but unseen forces from within the alien Cone could now exert an influence throughout The Hill.

No sooner was 'Ti'Elen on her own than it came to pass that the mysterious influence inspired a Mandarin to take 'Ti'Elen to a large foyer that was the last area the People's Orators passed through on their way into the Green Chamber. A balcony embraced all four sides of this foyer and it was enclosed not by a railing but by a parapet of solid stone. The ceiling of the foyer was composed of a lacework of stone frames, like windows, and deeply inset in the frames were pictures on glass depicting ships and trains and harvesters, the mechanisms of trade and industry and agriculture.

But the Mandarin indicated the vast expanse of naked stone that formed the balcony parapet, and he said to 'Ti'Elen, "This is for you."

And he took her into the Green Chamber and he showed her tall, graceful Gothic windows filled with indifferent glass and said, "These, too, are for you."

And between the windows great blocks of uncarved stone protruded nakedly from the limestone walls. "All this," said the Mandarin, "these blocks, the windows, and the foyer parapet – these are for you."

And 'Ti'Elen's mind staggered with wonder and amazement.

She had come to the heart of The Hill and The Hill had no heart. It was still blank.

She knew that a country is more than the machines of trade and industry depicted on the foyer ceiling. She knew that all the leaves and vines and animals and heads elsewhere amounted to nothing without a common memory and the promise of a common future. True, the astoundingly beautiful memorial chamber in the Great Tower, with its altar and its angels and images of battle, displayed a selective memory of the valorous deeds of war, but where was the context? The heart was blank.

That blank heart was now her dominion.

It was up to 'Ti'Elen to carve history. She was to carve Canada.

CREATION

PHASE I

Old One repeated, "The stone heart of The Hill was blank, and 'Ti'Elen was to carve Canada."

"You can't carve Canada," said Youth. "Canada is a country."

"*Is*," said Old One, "but how did it begin? Where did it come from?"

"It just happened," said Boy.

"Nothing just happens," said Girl.

Old One agreed. "Life, Death, stones, trees, Love, Hate – everything is created. In the beginning there is Mind. First the Thought, then the Deed."

"Of course!" said Girl. "Michelangelo could imagine the shapes in the stone. He just released them. I'll bet 'Ti'Elen carved what she imagined in the stone."

Old One chuckled. "And old stone it was, too. Ancient stone. Stone from the roots of the earth. What stories it told her. And 'Ti'Elen listened. But more than that," – Old One was almost whispering – "she carved what she *thought*."

"Tell us," they said.

And Old One continued.

*****

The great stone Cone sat silently on the cliff edge that formed the back of The Hill. It was attached to The Building, part of it but aloof, joined but separate, sheltered but exposed, a wondrous vessel attached to its terminal.

At first, 'Ti'Elen may not have fully realized the importance of the Cone but she knew she had to study, and research, and think. Study and research are not always attributed to artists. We think they create entirely from air and inspiration, and – Girl interrupted. "Michelangelo studied anatomy."

"What's an atomee?" asked Boy.

"He cut up dead bodies to understand the muscles."

"Yuk!" said Boy.

Old One smiled, and continued.

'Ti'Elen had to study the anatomy of a country. What better place to begin than in the mysterious Cone that sheltered memories and dreams. Soon, guided by the impulse to learn, 'Ti'Elen found herself walking down the Hall of Honour that was still almost bare of honour, and approaching the massive doors that blocked the now unguarded entrance to the Cone. She felt impelled to approach, ordered from within and called from beyond. With beating heart she depressed the great latch that released the double door. One side swung wide with amazing ease as though the powers within were awaiting her arrival.

She entered the Cone and the door slammed behind her.

Her breath froze in her face. Her hands turned numb. Her eyelashes froze to her cheeks. She gasped for air and at first saw nothing but a mist of frozen crystals.

Oh, the agony of it! The searing pain of the cold. The torture of trying to see through a whiteness that was so intense she could touch it.

Then a finger of wind brushed lightly by and began to draw pictures in the ice crystals like a pencil of frost sketching on frozen window glass. And she moved forward and into the frost and became one with the ice-atom pictures and was no longer cold.

What amazing things she saw! It was as though she were standing upon the northern half of a continent, but not upon grass, or soil, or even rock. She was standing on ice. Ice a mile deep. Ice so thick and heavy that the sheer weight of it caused its bottom layers to flow slowly southward, gouging valleys as they went, grinding the eastern mountains that had been heaved up in the world's beginnings, grinding them until they became rounded hills. She saw western crags standing jagged against the sky and in this vision she saw a strange Corridor like a great hallway left miraculously between mountains and ice sheets, a Corridor that ran from north to south. At the top of the Corridor, as if by magic, a bridge led from Asia and across this bridge she saw all manner of animals moving to the Corridor and down into the continent.

'Ti'Elen peered into the Corridor and cried out with alarm. "This is not a country! This is a wilderness of animals locked between mountains and surrounded by frozen nothingness."

And she felt voices inside her head. It was as though Essence of Mind were speaking to her, an essence that was all about her, incorporated into the frozen air that occupied the mysterious Cone. And Essence of Mind used a voice that was many voices.

Many Voice whistled and moaned as though the sound came from the throat of the North Wind that slid forever across glaciated ages. "What is a country?"

it moaned. "Is it only an area within a border?" Many Voice seemed to laugh mockingly.

"Is it a collection of buildings within a fence?" There was a tone of scorn in the vibrating air, and the wind funnelled a vortex of snow that rushed upon her and towered before her and 'Ti'Elen trembled with fear.

"Is a country an organization?" howled Many Voice. "Is it an in-sti-tu-tion?" The multiple voice rose upward and then laughed in derision as 'Ti'Elen hesitated, and knew she must answer these questions or forever abandon her task.

"No," she quavered, her own voice barely more than a whisper. "A country is people."

"Is that all?" bellowed Many Voice through the wind.

Suddenly 'Ti'Elen grew angry and shouted back. "All right, then, it's an idea!"

"Oh?" and now it was more a question than a challenge.

"Yes," she said, standing as tall as her full six feet would allow, as though to outface the awesome vortex that still whirled before her, "and an idea begins with people!"

And the vortex melted and the wind sighed and warm air caressed her face.

"Very well," said Mind, speaking softly through Many Voice, "begin with people."

"I intend to," she answered.

The ice turned to oceans, the oceans to land, the wind to birds, and Mind wafted her to her drawing board in the basement studio where she seized pencil and paper and began the task of creation.

She imagined Cave Men migrating across the great ice bridge and down the Corridor in search of food.

And she took mallet and chisel and went to the foyer of the Green Chamber. There on a column of stone embedded in the balcony parapet she carved a  family of cave dwellers, primitive creatures not yet fully human but wielding crude tools and nurturing children. And so it was that long before there was Canada there was Man, Woman, and Child. And since Man, Woman, and Child and all else that was to follow had roots in a time that was long before Gothic, 'Ti'Elen used a primitive style that sculptors called Romanesque. Romanesque had the simple, rugged look of unfinished work, which suited the unfinished nature of humankind.

And she imagined other newcomers exploring across the great bridge at the close of an Ice Age. She saw, in her mind's eye, highly evolved peoples, *homo sapiens* these, cleverly adapting to northern climes.

'Ti'Elen made designs on paper, sketches that sculptors call cartoons. She spread them on the floor and then, standing on the scaffold, drew them in charcoal on the stone. So accurate was her hand that the transfer could not have been more perfect had it been projected and traced.

Now it so happened that the old Master Carver had left so quickly he had given 'Ti'Elen no advice or instructions, and so there were things she did not know. She did not know that as "boss" she had to sign certain cards before her men could be paid. The men waited patiently for several weeks, then a spokesman humbly informed her of the necessary ritual.

'Ti'Elen had never used an air-driven chisel. She asked the men to teach her, and they did, but they complained that it was not work for a lady.

She did not know what the men were capable of doing and in those days of the Beginnings she kept them busy cleaning stone with air hammers until she could find out. But she could not hope to do everything by herself. Even the great and famous Rodin had seldom carved the finished work himself. He made models from which assistants made the full-scale figures.

There was a carver from Denmark, Anton Nielson. He had laboured as a slave during the second worldwide war and was now aging and in poor health. 'Ti'Elen tested Anton by letting him carve the fur of the cave family's clothing. She was very pleased, and as time went on Anton taught 'Ti'Elen a great deal about stone. About how it was grained and how it reacted to the steel and how far it could be forced to reveal shapes.

And there was a carver from Italy, Iti Busolo, who had worked with a master sculptor in the old country. Soon she was letting Iti remove the raw stone that had to be taken away from around the sketches in order to reveal the presence of figures. She would mark on the sketches how deep to go and he would follow these instructions. This was called roughing out and it meant that 'Ti'Elen could confine her strength and her skill to releasing the figures themselves. Sometimes Iti liked to play tricks and would leave her a challenge by taking away too much stone, and she would change a figure in subtle ways to compensate.

And so it was that, as the mallets swung and the chisels sang, an Inuit hunter clothed in mukluks and swirling furs emerged from the stone, caught in the process of teaching his son to hunt.

And she carved smooth-skinned Indian hunters, working in unison, spears poised, points aligned. It was a vision of people who were capable of creating intricately organized social structures while living in a world in which Nature and Spirit were one.

Should they, too, come across the bridge and down the Corridor? Or would they cross the Pacific in

sail-winged canoes and migrate up from the south, hunting the Mastodon as it followed the retreating ice sheets even into the rocky wilderness where The Hill itself now stood?

Whatever the route, she carved and they came.

"I will imagine," said 'Ti'Elen to herself, seeing pictures in the back of her head, "only happenings that, occurring even once, will guide the course of history."

She went from drawing board to Cone and back to drawing board. She saw visions of graceful ships and of tall fair-haired men, mighty-muscled, steel-armed, wearing horned helmets. "It is a thousand years ago," she thought, "but I see these white-skin warrior sailors being greeted by the red-skin hunters who already inhabit the land." She saw visions of sod-domed dwellings by the eastern sea as the first men to arrive from Europe sought to establish homes. But what she sketched and carved were men, Viking and Indian, the white and the red, their futures forever entwined.

CREATION 33

Anton the carver had Viking blood in his veins and it was Anton who carved the horned helmets and who turned the high-prowed ships of *'Ti'Elen's* imaginings into a ship in stone.

So it was carved and thus the Vikings came.

You must understand that this work of creation that was underway just outside the doors of the Green Chamber was not the happening of a few days. A country is not created overnight. It was, however, created at night. Late in the evenings, after the People's Orators had departed the Green Chamber, 'Ti'Elen and her crew came up from their sub-basement workshop, so deep in The Building that few people knew it was there, and took over the foyer. They erected scaffolding there, and hung it about with heavy canvas. They had chisels and mallets but they also had powerful pneumatic tools driven by compressed air, and the noise of their work made The Building rumble like an old man's snore, and the dust of creation hung as heavy as the smoke of the Firebird's fire when the world was begun.

They were a friendly crew, these stone carvers, though all were not happy to be ruled by a woman. But mostly they were cheerful. They brought their own food for late night meals and they ate it while sitting on the staircase where one day a prime minister would pause in order to speak in a lordly fashion to journalistic cameras. They played practical jokes that masked the momentous nature of their work, such as painting boots and filling a lunch-box banana with plaster of Paris. Occasionally they drank too much before coming to work, an error that also afflicted many Bureaucrats as well as many of the People's Orators and even the aged inhabitants of the otherworldly Red Chamber. Iti Busolo, being from Italy, brought lunchtime wine in a jam jar but that was considered culturally correct. And some of the others knew of hidden beer caches in cleaners' closets. 'Ti'Elen occasionally noticed that her crew became more mellow during the second half of the night.

But 'Ti'Elen understood the human frailties.

She had difficulty, however, understanding the men. For all that they were for the most part conscientious and polite, even friendly, there seemed to be a barrier, a gulf between them and her. One day she took courage and spoke up.

"I don't understand you men. You don't trust me. Why not? We work together. I give you challenging work. I take you to your home if you're hung over. I help your wives when they're sick. We play cards. Your children call me 'aunt.' I have even lived in one of your homes."

Piece by piece, the answer came to her. These men, skilled in their craft and with years of experience, had been treated like mere labourers. They had been underpaid, and laid off, and re-hired almost at whim. They had done the Master Carver's bidding, be it smoothing Cleophas Soucy's work or painting grooves or whatever else he ordered them to do.

But now, a Bureaucrat saw what 'Ti'Elen was creating and felt ashamed to be rewarding an artist with a tradesman's wages. He wished to increase her pay. 'Ti'Elen accepted only on condition he would also raise the carvers' wages. It was done.

Even so, Iti Busolo told her one day, "You are not a good boss!"

"Why?" asked 'Ti'Elen.

"You don't tell us when our work is good."

And 'Ti'Elen thought about it. She realized that her own parents had seldom criticized but had never complimented and that even she missed the encouragement of justified praise.

As months went by 'Ti'Elen learned what the men could do and learned what the men thought and together they carved. She tried to let each carve according to his talents. There was Joe Joanisse, he who had nicknamed her 'Ti'Elen. There was Anton Nielson, Iti Busolo, Wilfrid Filion, Fernand Rossignol, and apprentices who came and went.

'Ti'Elen would commune with Essence of Mind and listen to Many Voice. She would turn her visions into sketches, transfer the sketches to the stone and then, late into the night, girdled by darkness, cloaked by canvas, they would toil at the rock. Each morning the Bureaucrats and the Orators were in possession of more of a country than they had possessed the day before.

"The Vikings were sailors," thought 'Ti'Elen. "Who will come after the Vikings if not other sailors?"

She imagined sailors who were not explorers but who were fishermen. They were common fishermen from Brittany who heard of the Viking adventures and sailed to the faraway underwater banks where fish lived in their millions. She saw these fishermen in her mind's eye but she did not carve, for they were secretively following where the Vikings had led. They were sailing under oath not to reveal the origin of their wonderful catches, and history was blind.

She saw an arrogant sea captain who was not blind. He was Giovanni Caboto of Italy, sailing under England's flag as John Cabot.

She imagined his ship, driven ahead by full-bellied sails but being held from shore by the sheer force of fish. Cabot grew under 'Ti'Elen's chisel and so did his serpent-prowed ship. Carver Wilfrid Filion drew fish from the depths of the stone and arranged them in unyielding opposition to the encroaching vessel.

Together they carved Business and Nature beginning eternal conflict in North America.

And so it was that John Cabot reported back to the businessmen of England that they should be paying more attention to what the fishermen already knew, that here was a source of wealth. The fishermen's trade secret was broken. Business had truly arrived in the waters of Canada.

While fishermen from Portugal, Spain, France and England swarmed to the coasts of what Cabot falsely claimed to be "new founde lande," 'Ti'Elen let her imagination roam ahead another third of a century and she saw another ship and another sailor.

She had visions of Jacques Cartier, a fisher captain from the seaport of St. Malo. She imagined him crossing the wide ocean as an agent of the King of France to "discover" land that the captain already knew would be there, for as a fisherman he had already met the natives and had heard their stories of a great river to the west and of rocks that gleamed in sunlight.

Cartier, she thought, would be searching for a vast waterway leading to wealthy kingdoms and the wealth sought by his patron king would not be fish, but gold.

She saw him finding two passageways past Cabot's "new founde lande," both leading to the same giant river streaming from inland, and that river again dividing as yet another fed it. The natives would tell him of a waterway so long that no man had ever seen the end of it.

She imagined Cartier, already bludgeoned by the heat of summer and almost destroyed by the cold of winter, having his own fearful visions of the unknown. With hammer, chisel and drill she carved him standing with one foot on shore, straining to see and to record what he could never see, for, being a sailor, his other foot was firmly planted on board ship. Around the ship multiple moons floated on the invisible flowing tide of Time.

That is how Jacques Cartier came to Canada.

You must understand that 'Ti'Elen's Cartier, as she carved him, came not only from France but also from

Yugoslavia. At university she had studied sculpture under Ivan Mestrovic, an artist revered in his native Yugoslavia and widely acclaimed abroad. Mestrovic created large, rugged, strong figures powerfully braced on strong legs. 'Ti'Elen had learned from a master about the drama and power of stance.

Cartier charted two entrance ways to the massive gulf that formed the mouth of the great river. He claimed all surrounding territory on behalf of his king, a quaint habit the Europeans had of claiming land already inhabited by others. He did not find the gold he sought, although it was there. But he found wealth in the form of the fur pelts the Indians exchanged for iron knives and utensils. With Jacques Cartier the fur trade came to the shores of Canada.

'Ti'Elen thought of the river and imagined it as a water highway pouring out of the unknown. She conveyed her vision to Fernand Rossignol, whose chisel had been creating luxuriant vegetation around Cartier's foot and rippling water beneath the keel of his ship. Inspired by 'Ti'Elen's vision Fernand's tools began to sing like his namesake the nightingale and he spiralled the turbulent waters of wild rivers down a stone column and put a ship on their waves and behold! – it came to pass that the white men journeyed into the interior.

'Ti'Elen's inner eye saw the first of those inland explorers. He was Samuel de Champlain. She imagined him leaving ships and sea far behind and venturing inland. She saw him as a man of courage, understanding, and deep religious faith. She drew him as a strong figure relating on equal terms with another strong figure, an Indian. In her vision Champlain

pointed as though to ask, "What lies yonder?" and the Indian put a hand on the European's shoulder as though to say, "Come, I will show you."

Behind Champlain and his companion her charcoal danced on the stone and trees appeared, as though the men were in an orchard. It was 'Ti'Elen's way of saying that she saw the Indians with a sophisticated culture of their own that was rooted and flourishing before the white men arrived and that Champlain would be permitted to move within it. And Iti hewed

away the stone in chunks and 'Ti'Elen struck it away in delicate pieces until the culture and the figures emerged. And it came to pass that Champlain moved far inland, north of the waters of the great river and the great lakes, and made alliances. He organized the fur trade.

It came to pass that Champlain fathered New France.

*****

Old One rested, eyes unfocussed, looking into the approaching past.

"What happened then?" asked Girl.

"Everything."

"Like what?" Youth was sceptical of ancient myths and Old One's tales.

"Like battles, adventures, explorations, conversions, baptisms, settlements, plagues, massacres, heroics, martyrdoms – all the things one would expect when proud races and different cultures meet, mingle, and do business."

"And I suppose," said Youth sarcastically, "that 'Ti'Elen created all that?"

"Not at all."

"Then why speak of it?"

"Because Girl asked."

"I asked," said Girl, "because you said that 'Ti'Elen carved and it came to pass."

"I said she carved the magic moment, that tiny event or period in time after which nothing would ever be the same again."

"Sounds to me," said Girl, "you're saying she shaped the future."

"We all shape the future."

"Hold on," said Youth. "What you've really said is that she shaped the past! That's impossible."

"There is no past," said Old One. "Only memory."

"Memory can be wrong."

"Indeed it can," said Old One, smiling warmly at Youth, "and you are wise to believe so. But out of memory grows mythology and out of mythology grows the essence of a country. No country has ever survived without a foundation of mythology. That is a paradox of history."

"But who lays the foundation?"

"Artists, of course. So you see, artists working in the Present can shape the Past and in doing so help shape the Future."

Youth, who had only recently discovered the pleasure of having his mind stretched by a paradox, remained silent but thoughtful.

"Never mind all that," said Boy, bored. "How'd they carve the stone?"

"Silly," said Girl. "With tools, of course."

"Hammers and chisels is all I've heard. And air-driven chisels. Sounds crude."

Old One sighed. "Crude and delicate. Follow the charcoal line with a fine- pointed scriber. Embed the line in the stone. Remove large unwanted areas with a clattering air-driven chisel. Move in closer with a purring chisel. Tap with a gentle mallet. Use drills. Use files, rasps, rifflers, call them what you will. Use broad chisels, narrow chisels, chisels with or without teeth. There is no law that regulates. The artist uses what the artist needs. Some say it's heresy to use power-driven tools. 'Ti'Elen maintained that just as the tool itself is an extension of the human hand, air power is merely an extension of human muscle. She used what was appropriate, her only concern being that power didn't detract from human touch or injure human muscles."

"She could draw the outline," said Youth, "but a sculptor works in depth. How do you draw depth? How do you sketch volume?"

Old One smiled. "You don't."

"You see it," said Girl.

"Some do," said Old One. "I think Fernand saw it. He was a master at undercutting. He could carve a leaf then go in behind it so deeply it would all but hang in the air. 'Ti'Elen recognized this talent in him and encouraged it and made use of it. She herself saw with her fingers."

"Next," complained Youth, "you'll tell us she was blind!"

Girl, Boy, and Youth waited for an explanation.

"'Ti'Elen had a gift for design and it may have been because she saw differently than you and I see. She could read words that were upside down, or backwards. When she was young she would sometimes see things as backwards that weren't backwards. Later, as an artist, it meant she could always see in ways that were extraordinary. And when it came to volume, she didn't see it, she felt it. Her hands would run over a sculptured arm and her fingers would tell her, this arm needs bulk. Her fingers would slide into the cavity behind and would tell her more stone must be removed."

The listeners tried to understand.

"I do not explain," said Old One, "I merely tell. And it's time now to move on down the wall, for most of the stone was still blank. New France had emerged but Canada was still unformed. What should 'Ti'Elen carve next? What would Essence of Mind direct her to create? I remind you that Thought was still rippling out of the cone-shaped tower that dominated the heights at the back of The Building. Without the psychic Presence within that curved pyramid all memory and vision would be erased. Each day would be isolated from its past and indifferent to its future. But the Cone was there and 'Ti'Elen was there, and all else followed."

Old One untied a small pouch and opened it. There were several compartments, each containing a powder.

"What are those?" asked Youth.

"In good time," came the enigmatic reply, and then, relenting, "Each is a source of a different beauty."

No more was said, and Youth held his peace. Old One would move at Old One's pace.

All three watched as Old One selected powder from one compartment and, leaning forward, sprinkled it on the fire. The flames exploded into a dance of so many colours that Girl thought for a moment Old One had captured the sunset. And, indeed, the sunset glow had gone from the sky. They were alone, four people in a circle, seated on the ribs of the World, guarded by Moonship and watched over by all the Star Eyes of heaven, and at the centre of their circle was a small fire and at the centre of the fire was Beauty.

All four watched the Flame Dance until it subsided, and then Old One continued.

CREATION

PHASE 2

The first few times 'Ti'Elen entered the Cone she did so in fear, her heart fluttering within her. But soon she went joyously, expecting the unexpected, excited by the unknown.

And always ideas hung thick as mist, and pictures whirled out of nothing into nothing, and sometimes she heard sounds of women singing and of drums softly beating. And other times she heard martial music and the stamp of booted feet. She heard the laughing loon and the sad owl and the gentle swish of dripping paddles.

Essence of Mind urged her forward along mental pathways where Past, Present, and Future were channelled together into the strange maze of Life. She saw images as clear as painted pictures.

She saw two great European powers engaging in commercial and military combat in many areas of the world. She saw the men of New France being used by the King of France to establish forts down the Ohio Valley, thereby irritating British colonists to the east. She saw British traders establishing posts in Hudson's Bay, thereby diverting the fur trade away from New France. She saw competition, turmoil, bloodshed, raids, burnings, retaliations, barbarities and heroics.

"This must end," said Mind to her mind, whispering through the lips of Many Voice.

And 'Ti'Elen fled from the Cone and took refuge in the deepest basement of The Building where she seized paper and pencil and sketched mightily. And she came up at night and sketched again upon the rock. And again the sound of steel on stone rang throughout The Building.

Hands reached up from concealing foliage, fingers straining for a hold.

Broad-shouldered men carrying muskets drew themselves up over the lip of a rocky plateau, muscled arms reaching forward, veined hands grasping at rooted plants.

'Ti'Elen could feel the climbers' fear as they struggled upward, for she, too, was a climber of rocks. With rope, spikes and climbing boots she liked to venture with others up the vertical faces of cliffs. She liked the challenge of the rock, the test of endurance and courage, the discipline of fear, and the enormous exhilaration of reaching the top and standing, free at last, to view the panorama below. Now she imagined nervous soldiers climbing upward through the blindness of night.

And so it came about that an army crawled up the stone and grasped for a hold on a level plain that spread its fertile carpet to the gates of a stone fortress.

CREATION 49

And so it came about that the British captured Quebec, the capital of New France, and history went in a new direction.

But all the time Many Voice whispered to 'Ti'Elen, saying "Context, context, think of context. These machinations of Man are merely the wild struggles of one species."

"Yes," thought 'Ti'Elen, "all Nature does not change direction because men fight a battle. A country may begin with people but there is more to a country than people."

And she carved a deer at rest beneath a spreading tree and around the tree

twined the vines of life. Within the shelter of the tree birds perched. The deer looked with curiosity toward the warriors but showed no apprehension. Men were at war but in spite of war Life flowed onward.

"Even men like peace," thought 'Ti'Elen, and she pictured the kings of Britain and France making peace and handing out pieces of a world they didn't own. She thought of them as being basically ignorant men, unaware of and uninterested in the people and the territories that were to grow into Canada but vitally interested in power, status, face, and wealth, and in politics as a means to all such ends.

So she carved them relaxed, almost friendly, as though having tea while discussing a treaty that was beyond their intellect to comprehend.

In this treaty the kings exchanged vast territories as casually as they themselves might have exchanged their cloaks.

And she carved on into the next panel, letting the lines flow smoothly, letting the figures remain seated and relaxed as more royal decrees were handed down from the thrones of power. It was the end of New France and the beginning of a British colony called Quebec.

At this stage in her act of creation 'Ti'Elen must have paused to think very carefully, for deep in the frightening chaos of the Cone her mind was seeing pictures of other colonies rising up in revolt. She saw revolution and violence as colonies to the south realigned themselves into a union of states. She heard the music of bold anthems and the ringing words of a declaration of independence that included an inspiring statement of human rights. But these were other visions, for others to carve. Her vision told her that a major effect of the southern revolution would be the consolidation of yet other British colonies – Nova Scotia, New Brunswick, Prince Edward Island, Ontario.

✼✼✼✼✼

Youth interrupted. "Did she carve all that?"

"No."

"Then she didn't carve Canada."

"I have told you," came the answer, Old One's voice smooth with patience, "she carved the turning points, the moments of crisis, or of new direction. The die was cast when the British stormed Quebec. All else followed."

"I have read," said Youth defiantly, "that Canada was created by the American Revolution."

Old One flared. "What you have probably read is that Canada was the bastard offspring of the American Revolution. A mere by-blow of someone else's passions."

"Something like that," smirked Youth.

"Ah, but that is someone else's vision."

"Please ignore him," said Girl. "What else did she see?"

"Why, she saw what she had seen when the Creating had begun, and would continue to see. She saw a family – Mother, Father, and Child."

"Do go on," said Girl, and Old One did.

✼✼✼✼✼

So 'Ti'Elen carved the family she would never have. She carved them relaxed and at peace, the strong man resting one hand on the shaft of his woodsman's axe and sheltering his son with the other. The boy held a pet in his own arms, and behind his father stood his mother, the woman equal in stature to the man, her right hand resting on her husband's axe-hand with a touch that simultaneously comforted and guided.

And so it came to pass that settlers poured into the land, many fleeing the revolution in the south, and for many years they built homes, cleared land, grew crops, hunted food, trapped fur, carried on trade, and raised children, all without the need to carry muskets to the fields.

But 'Ti'Elen was always drawn back into the mysterious recesses of the Cone.

And now she caught tremors of violence and conflict. In her imagination she saw cannon belching fire, homes burning, Indian braves and white soldiers dying side by side.

Once again her charcoal flew over the unscarred stone and once again an assistant's large chisel tore the stone away in chunks, preparing the way for 'Ti'Elen. And this time the foliage that grew under the carver's touch was tortured and writhing and the figures shaped by the sculptor were no longer at rest.

A proud horse danced in anticipation of battle and from its back a tall man shouted orders. Armed soldiers charged forward. At the forefront of the charge a red man and a white man ran shoulder to shoulder, muskets firing, smoke wreathing upward.

And so it was that Canadians reacted fiercely to an invasion from the country to the south.

There were many battles, and General Brock, the tall rider of the tall horse, was killed, but his spirit lived on. And Tecumseh, the Indian war chief, was killed, but his spirit, too, lived on. And the invaders were repelled. And the borders of Canada, from the edges of the Atlantic Ocean to the innermost tip of the inland seas, were defined. Under the hammering blades Canada was taking shape, but mingled with the dust of carving was the smoke of guns.

*****

"Gee!" said Boy.

"So much for the undefended border," said Youth.

From out on the lake came the long mournful wail of a loon and from the woods behind came the sudden scream of a rabbit dying in an owl's sudden grasp.

Old One and the Listeners sat staring at a fresh log on the small fire where ants scurried for safety and were consumed. The flames still bannered in beauty but at their heart was Death.

CREATION

PHASE 3

'Ti'Elen prowled the recesses of the great Cone and found a miasma of spiritual confusion. It was as though waves of darkness were rolling through waves of light. Hidden in the dark corners were yawning pits reeking with the foul smell of ignorance. She groped from darkness into blinding glare and found verdant vines of knowledge spiralling upward around pillars of sparkling insight. There were smells of decay that made the senses reel and yet swelling out of the putrefaction came all the sweet perfumes of life. Winged spirits clothed in black beat invisible paths through hateful clouds, poisoned darts of malice shooting from malevolent eyes, while other shimmering spirits danced lovingly in an azure dawn and the sounds of sweet song filled the air.

'Ti'Elen knew not where she was or what she saw. Was she was in a cauldron where Good and Evil were mixed by the swirling currents of life? Were the voices singing the sweet music of Paradise or the sad songs of Perdition? Was the Evil around her what had been or what would be? Was the Good enveloping her the Good to come or the Good that would be no more?

She fled for sanctuary to the foot of a tall spire of white marble that rose wraith-like at the core of the Cone and here she heard Many Voice murmuring to her. There was only one word and it was intoned softly, repeated, like a chant. "Religion, Religion, Religion, Religion – "

And 'Ti'Elen remembered that she had created Champlain, a religious but ambitious man, and that much else had followed. She knew what she must do, and she seized chalk and chisel and returned to the endless wall. This time she took carver Joe Joanisse with her, he who had nicknamed her 'Ti'Elen. And again The Hill echoed to the song of the chisel as the chips fell and the dust billowed, and lo! a cleric in a long robe and a high collar sat by a Tree of Life, Bible in one hand, Cross in the other, and spoke of God revealed in Christ. In her carving he spoke to people who already knew of the Great Spirit through other revelations and other symbols.

Christianity had come to Canada, but more than that – organized religion had arrived.

Old One fell silent, lost in reverie.

*****

"What are you saying?" asked Youth. "Was that the Good? Or was that the Bad? 'Ti'Elen felt all that rot and stink before she carved!"

"Oh, no!" said Girl. "She saw beauty and light and spirits dancing."

"Come off it," said Boy. "She was just dreaming."

"Perhaps," said Old One. "Or was one the Past and one the Future, and if so, which was which?"

"That's simple," said Boy. "All that darkness was the past."

Old One looked at Girl. She shook her head. The ancient eyes turned inquiringly toward Youth.

"I don't know. The old religion bad? The new religion good? I think there's always some of both."

Old One smiled.

"What I want to know," said Boy, "who was the guy in the long robe? Was he a Catholic or a Protestant?"

"Ah," said Old One, "'Ti'Elen did not specify. The priest she carved held a cross, but the cross carried no body of Christ. Perhaps on purpose?"

"On purpose!" exclaimed Boy.

"'Ti'Elen's purpose. Some crosses display the body of Christ. Some do not. With a body, the symbol is usually Roman Catholic. Without one, the symbol could be Protestant."

"Usually? Could be? Which was he?" asked Boy who liked straight answers.

Girl poked Boy. "Don't be silly. The Catholics came first."

Boy elbowed Girl.

Old One sighed. "I will tell you a story. Two priests came ashore with Champlain. One was Catholic, one Protestant. Those two priests were always arguing. They argued about Salvation, and Grace, and Faith, and Good Works, and Papal Infallibility. Like two dogs with opposite ends of one bone they snarled and snapped and wrestled. It so happened they died at the same time of the same disease. The sailors and soldiers, overcome with curiosity, buried them in one grave to see if they would fight in the Hereafter."

"Did they?" asked Boy.

Old One laughed. "Probably."

"'Ti'Elen – " said Girl, " – was she religious?"

"A questionable word. She believed in a loving God and a resurrected Christ and, in a way, in Heaven and Hell. Her first commission as a sculptor was to carve a Fatima group – three Spanish children who claimed to have had visions of the Mother of Christ. 'Ti'Elen didn't believe in such visions but as a sculptor she took great pleasure in exploring spiritual mystery. She carved them in wood. Simple figures, rooted in faith, striving to see the unseeable."

Youth was intrigued. "But she didn't believe?"

"An artist," said Old One, "doesn't have to believe what others believe to understand their yearnings. Or the inner Truth." He added, with a twinkle, " – besides, that first commission helped pay for her studies."

Youth smiled. "Okay then, having carved religion into Canada, what did she carve next?"

Old One's eyes squinted as though peering into distance past a bright sun, and Old One's voice resonated with wonder.

*****

In the panel that brought religion there was a tree, bearing fruit. It grew up behind the shoulder of the priest. Soon, Fernand Rossignol's tools carried that tree over to a pillar of stone and the branches and the fruit twined around it. In front of the tree 'Ti'Elen carved a figure of a tall youth holding an apple, and at his side a child reaching to pluck an apple and at his feet another child cradling an open book. Here was the Tree of Knowledge, the fruit of which caused Adam and Eve to leave Paradise, the very fruit that would always entice humans to reach beyond reach, to question the unanswerable, and to strive for the unattainable.

And then, beyond the Tree of Knowledge, the chisels of 'Ti'Elen and Iti Busolo carried on, hammering 'Ti'Elen's visions into the long wall. Out of the stone a teacher appeared, with children seated at her feet.

And so more knowledge came to Canada, and teachers, and schools, for who can imagine a country without teachers, and without children and schools?

"Some day," said Old One, scanning the Listeners, "you will visit The Hill. Look at the faces of the children on the western wall and you'll see how 'Ti'Elen felt about teachers and learning, and you'll know why she carved education into Canada. She needed no impulse from the Cone."

*****

"Then the Cone," teased Youth, "is a Story Teller's frill?"

"Not at all. But she knew that she herself was a product of education. The artist is not just a pair of unskilled hands waiting to be propelled by inspiration." Old One smiled. "But inspiration drives the skilled and sustains the weary."

Old One stared off a moment as though seeing beyond a far dark horizon. A sigh preceded words.

"And oh the country 'Ti'Elen was creating was bewilderingly vast, strange, and unknown. The years were rolling by and she was becoming tired. So were the carvers who worked with her, Fernand Rossignol, Anton Nielson, Iti Busolo, and Joe Joanisse."

Old One leaned forward and peered intently at the Listeners. The ancient voice dropped to a whisper. "But the psychic Powers within the Cone reached out to her, so once again she ventured into its depths in search of inspiration."

*****

CREATION
PHASE 4

Within the sanctuary of the Cone all was tranquil. 'Ti'Elen wandered at first as though she was in a dense forest, and then she was on open land and a wide river barred her way and this river ran from the north to the south. Beyond it, rolling flatlands covered with waving grass stretched to the edge of infinity and she knew she was seeing what was and had been, but there was no glimpse of what would be.

And then, as she stood, nervous, in a silence so vast it smothered the universe, she became aware of vibrations and of small sounds as though the ghosts of ground-nesting birds were taking flight. Vibrations increased and sounds increased. Invisible feet were heard scurrying as though small creatures were seeking sanctuary and she felt on her face the wind of wings. The sound of sharp hooves floated by as though antelope, too, were in bounding flight. The ground began to shake and the thrum of wings and the thump of small hooves became submerged in a louder roar that rolled toward her like summer thunder. She thought of an avalanche in the high Himalayas but all around was level as the hand of God. And suddenly she was enveloped in dust and she fell, choking, and the sound of a million hooves rolled by, around, and over her and in the midst of the thunder she heard horses snorting and men shouting and rifles firing and the sliding whine of arrows.

She lay, face down, as the sound of the hoofed hordes dwindled and the dust hung heavy as a shroud. From far in the dense distance she heard yet another sound, low at first, but approaching, and it sounded more fearsome than any she had yet heard. It was the cry of bagpipes lamenting lost homes in faraway lands, and the wail of the pipes made tears come to her eyes. It was the skirl of bagpipes promising new homes in a new land, and the song of the pipes made her heart beat faster and her blood run hot.

And 'Ti'Elen knew what had been and what must be and again she took chalk and chisel and returned to the wall.

And again trees grew under the chisel, for Nature is always present, and this time heavy grasses abounded, and cutting across the panel from top to bottom ran a river. Standing amidst the grasses, beside the river, there appeared the massive figure of a settler, booted feet wide braced, sleeves rolled, arms swinging a double-bladed axe. Beside the river there were cultivated fields in which a woman stood. Within the folds of her apron she held seeds of the good earth and

beyond her a two-wheeled cart stood ready to carry produce to market and people to far frontiers.

It was a pastoral scene quite at odds with the turbulent sounds she had heard in the Cone, but the scene was carved and Scottish settlers came to the Red River Valley and much else followed.

Cultures came into conflict. The buffalo hunt and the wild freedoms of the prairie natives gave way to the demands of settlement and agriculture. The trade routes of warring fur empires were altered. 'Ti'Elen had carved the seeds that would grow into the boundless wheatfields of the western prairies, into ever-moving settlement, into massacres, rebellions and hangings, into subjugation for some and enormous freedom for others. 'Ti'Elen's chisel did not carve Good or Bad but laid the foundation for what would be.

She carved onward into a pillar of stone in which, in her mind's eye, she saw a man struggling through heavy undergrowth.

"This man," she thought, "is more than an explorer. He is a surveyor and a mapmaker. There may be explorers before him but he will not only cross great mountains to a western ocean, he will chart the rivers that flow through those mountains and confirm the shape of the mountain passes. He will chart more precisely than any man before him and to the wonderment of many who will follow."

Thus of all the explorers and surveyors who would make the maps that showed the way for settlers and businessmen to push Canada to the shores of the Pacific Ocean it was David Thompson who was created by the chisel of 'Ti'Elen. It was David Thompson, who would die neglected, forgotten, and in poverty, that 'Ti'Elen created in stone.

Throughout the long nights the marble floors and the limestone walls of The Building continued to echo to the muffled rumble of fist-sized air hammers driving broad chisels and to the high-pitched buzz of pencil air hammers and the tap tap of mallets. And always, in the morning, the dust had vanished and so had the phantom figures wearing smocks, coveralls, masks, goggles, and ear defenders. And always imagination had pushed the bounds of Canada farther afield than they had been before.

One morning the People's Orators and the Bureaucrats passed through the lobby on their way to their appointed tasks to find that a giant figure of a man had emerged from the stone.

He was booted and muscled and was standing with his arms spread, hands braced against mountain walls, mighty biceps straining. One could almost hear the very roots of the earth being torn loose as mountain barriers were forced apart. And at his feet a railroad ran and a steam-driven engine puffed its way toward the gap in the sundered mountains.

And so it was that the railroads, having passed through, nay, having created rebellions and turmoil in the Red River Valley, and having ribboned across the prairies carrying settlers whose presence would change forever the nature of the vast plains, now passed through the Rocky Mountains and on to the Pacific coast. The colony called British Columbia joined Canada and the provinces of Manitoba, Saskatchewan, and Alberta were soon formed. And so it was that Canada, by spreading from the Atlantic to the Pacific, cut off the aspirations of the country to the south that had hoped to expand north to the Arctic snows.

And deep in the recesses of the Cone the Arctic snows were again calling 'Ti'Elen, but these were no longer the primordial glaciers of her first encounters. Essence of Mind was giving her strange impressions that were void of both image and sound and yet seemed to imply there was a vital ingredient she must still incorporate into her visions if a whole country was to emerge.

She searched for this vital ingredient but lost it in a blinding surge of wind-driven snow, and through the blizzard she saw the forms of dogs running in a harness that let them run widespread, fanned out, each dog pulling directly on the following sleigh. And behind the sleigh ran a gaunt man, wind-weathered and determined. She had come full circle from her first visions but everything was different.

'Ti'Elen went to her drawing board and sketched men of different cultures exchanging trade goods, to reaffirm what had already been established, that without commerce there can be no country.

And she puzzled over the vital ingredient that she had sensed but not seen and finally she sketched the figure of a banker, for without control of finance there can also be no country.

    Her pencil strove onward and soon new visions were being charcoaled onto the stone. And then, disaster struck.

    'Ti'Elen, the woman who was carving Canada, was almost killed.

*****

Old One sat quietly for a moment, lost either in reverie or sudden slumber. The Listeners waited. On the sloping rock by the starlit lake the small fire was burning low. Old One stirred, and plucked a small wildflower from its shelter in a rock crevice. The pouch was again opened, and the old fingers carefully sprinkled the little plant with another powder, and dropped the plant into the waning flames. The fire flared up, giving both light and heat, but neither stalk, nor leaves, nor blooms were consumed. A little while ago, Death had been in the heart of the fire but now the same beauty preserved Life.

Old One continued to sit in silence, head bowed.

Eventually, Boy shuffled, Girl fidgeted, and Youth coughed.

Girl broke the silence. "What happened? Did she fall?"

"Ah," said Old One, startled. "Fall? Well might you ask. They were spending their lives on scaffolding, those carvers, working long hours in noise and dust. A step backward to admire one's work, or a slip – yes, yes, quite possible. Indeed, before 'Ti'Elen's time an experienced carver, working high in a corner of the very same lobby, had had a great stone come down upon him, crushing his chest. He was working alone, and he died. Since that time no carver worked alone."

"But what happened to 'Ti'Elen?" said Girl, becoming impatient. "It couldn't have been in The Building. Many Voice would have warned her. Essence of Mind would have protected her."

Youth snorted. "You believe all that?"

"I do," said Boy.

"Maybe not," said Girl, wistfully, then added defiantly, "but I understand it." She turned to Old One, almost pleading. "The Building wouldn't have hurt 'Ti'Elen?"

Old One smiled. "No, indeed."

"So what happened?" asked Youth.

"'Ti'Elen had a little car. Low slung. Fast. She loved that car. Every year she'd have the engine taken right apart and cleaned and reassembled and fine tuned."

"I don't think I want to hear this," said Youth. "I bet she liked to burn rubber."

"Indeed, but not this time. It was a lovely autumn day. She'd been up in the hills beyond the river, driving with the top down to enjoy the warm sun. She'd stopped at a farm and bought fresh eggs and fresh whipping cream. She'd taken shelter for ten minutes from a sudden warm cloudburst but was quite relaxed and feeling no pressure of time. In fact, she was driving slowly, heading back to the Town for a leisurely supper before going to work on The Hill.

"She was driving down a steep hill that had great twists at the bottom of it, bends so sharp it was called the Dog's Leg, when suddenly around a bend, approaching her, came a large car quite out of control. It was sliding and skidding on the wet road."

"A head-on!" cried Girl, horrified.

"Not at all. 'Ti'Elen was a skilled driver and she knew this road to perfection. She swerved abruptly into a driveway, the little car answering promptly and remaining glued to its tracks. She was completely off the road, in a driveway, but the big car came through the ditch and smashed into the side of her car. The door on the other side flew open and 'Ti'Elen was flung half out, where she hung, head down.

"She lay there for minutes that seemed like hours and wondered about a bright light that seemed to shine from the woods above her."

"The Cone," breathed Girl.

"Concussion," said Youth.

"I don't explain, I merely tell," said Old One, "but as she stared at the light a voice spoke to her which was quite possibly her own voice. It said, ''Ti'Elen, you're going to be all right. You haven't finished your work yet'."

"Many Voice!" said Girl.

Old One smiled but did not comment. "Next thing she knew she was in hospital. One hand and arm badly cut, glass pieces in her head and face, toes broken almost off, legs cut, knees crushed. But she recovered. Oh, her hair started to turn white – it was early for the snow to come. She walked with two canes for quite some time and after that had to give up rock climbing and skiing."

"But the carving," said Girl, "what happened to the carving?"

"Remember, she had prepared sketches of fur traders and voyageurs and of the figure of a banker. Anton Nielson and Fernand Rossignol, the Viking and the nightingale, carried on as best they could for the five months she was absent. But these panels and the figure suffered from a vagueness of vision. Panels of fur traders and voyageurs tended to reinforce what had already been established, that business reached into the farthest corners of Canada."

Old One chuckled. "And banking, although essential, is not an inspiring occupation.

"But, when the dust had settled, more had been created than trade. It came to pass that Canadians wishing to do business would often have to risk long journeys into the unknown. In such a vast country Canadians, like 'Ti'Elen's voyageurs venturing through Fernand's wild water, would often feel like strangers in their own land."

"But there was more," said Girl. "What of the North Pole and who was the determined man with the dogs?"

"Shall I continue?" asked Old One, and the three answered with the voice of one. "Yes."

*****

Canada now had within its own control the mechanisms of trade and finance. Without such control no state is a true country but merely a colony. But its northern boundaries were not yet defined. Its northern reaches seemed to disappear like the foredeck of a whaling ship vanishing into a blizzard.

'Ti'Elen's vision cleared and she strengthened her sketches and the Rossignol's chisel carved ice out of stone while 'Ti'Elen's blades conjured up a figure. Soon, from around a fur-clad figure great jagged shards of light radiated outward and the lights were sprinkled with stars and all were encompassed within a great bowl like the very bowl of Heaven. And the figure itself was planting a flag, but not to claim land, for here there was no land, only ice crowning the top of the world.

And so it was that men journeyed to the North Pole and the light that shone around them was the Aurora Borealis. But the light that radiated in the stone was also the light that shone upon 'Ti'Elen when the Voice spoke of work yet to be done.

The pounding chisels drove on into a column of stone and revealed the wind-weathered man of 'Ti'Elen's earlier vision. He was no longer guiding his sleigh but was standing, parka-clad, looking ahead at exhaustion, loneliness, privation, and achievement.

And so it was that Law and Order came to the Canadian wilds; that the North West Mounted Police rode west on horseback and sleighed into the Arctic with dogs, and Canada's dominion stretched even to the northern pole.

*****

Old One paused for breath and Girl broke in.

"Why would 'Ti'Elen show Law and Order as one lonely figure in a parka?"

"Ah, but anything else would have defied reason. Just a few million people were taking charge of a land mass larger than imagination could conceive. Even the northern lands were larger than the whole of Europe. So few people in such a vastness could never maintain Law and Order with marching armies and charging cavalry. It would have to depend upon individual will, personality, and consent. Here Justice could only flourish by consent. 'Ti'Elen understood this and carved the figure without a uniform. In her country, Law is the people and the people are the Law.

"So 'Ti'Elen carved the lonely figure and it came to pass that much of the story of Law and Order in Canada was a saga of lonely men administering a lonely justice. She also decreed in stone that upholding the Law would always be a task causing anguish, labour, and pain. But I remind you, the creator is always in the creation. That gaunt stone figure of anguish and pain also reflected 'Ti'Elen's own anguish."

"But if she was carving again, she must have recovered?" said Youth.

"Not completely. Physical recovery was very slow."

Girl was shocked. "Then why did they make her work?"

"She made herself work. She was afraid not to. She was afraid of what might happen if she had to quit."

"What would happen?" asked Boy.

Old One chose words carefully. "Everything might come to an end. She herself, the person, might come to an end."

"But," said Girl, puzzled, "if she worked too hard – she could die!"

"I didn't say she was afraid of dying."

"Then what was she afraid of?"

Old One played deaf and continued.

*****

'Ti'Elen worked ahead, fear hanging over her, and out of the stone grew a man on his knees, by a stream, panning for gold. So intent was he on finding the precious metal that he was unaware of danger, even though behind him a mountain lion crouched on a ledge, ready to spring, attracted by the scent of greed.

And so it came to pass that the presence of Law and Order made it possible for Canadians to venture far afield to turn the Earth's vast resources into gold. But it also came to pass that danger would always lurk behind those who turned their backs on Nature, while exploiting Nature.

*****

"You were going to answer a question," said Girl, "but instead you told more of the story."

"Question? What question?"

"'Ti'Elen's fear," said Girl.

Old One sighed. "How to explain? 'Ti'Elen had been injured. Desperately injured. She was recovering, but slowly. She had a vast amount of work still to do – carving a country is not easy, you know – and it was possible she couldn't make it. But what if she stopped? What if she quit?"

"The Future," said Girl, "would vanish."

"Yes," said Old One, "but more than that. Her sense of who she was, her inner identity, her aspirations – all could evaporate."

"Could? Or would?" asked Youth.

"Ah, that was the problem. Was it a possibility, or an inevitability? If merely a possibility, should she take the risk? She couldn't. Even the possibility was too frightening, too threatening. She forced herself back to work."

"What was frightening? What could destroy her?" exclaimed Youth, exasperated.

"Why, loneliness, humiliation, ambition thwarted, frustration, failure – life is no easier for an artist than for anyone else! She forced herself back to the stone. She worked through that terrible period aided by the men, but aided mostly by her own determination."

The Listeners could tell by the rhythms that Old One was story telling again and they fell silent.

*****

CREATION
PHASE 5

'Ti'Elen was almost afraid to return to the Cone. Her last adventure there had been vague, the images indistinct, the impressions nebulous.

Her initial task was almost completed. The carvers' chisels had ripped a country from the stone-ribbed Past and placed it in the Present. But there was still an expanse of blank stone ahead of her that had to be carved to complete the square of the lobby and to close the circle of history. What should be here? What could be here? What would sum up and define but not limit?

She knew that the answers to such questions could only come from within the Cone – from Essence of Mind and Many Voice. But what if she ventured again and found nothing?

Two stone panels and a column were still blank and it was her destiny to carve them. But to what end? History is never complete. How could she create an ending that would be no ending?

If she sought inspiration once more, would it inspire or frighten? What terrible visions might she see?

She had to seek but was afraid to find.

Finally she gathered her courage around her like a warm cape and once again ventured down the echoing Hall of Honour. She came to doors of native pine set beneath the high arched frame of stone. Again she depressed the latch and the heavy door swung silently open. Again she passed within and the door swung as silently closed.

For a long time she was within, and in the upper recesses of the outer hallway small Gothic creatures watched with stone eyes and waited with stone hearts for her safe return.

When she emerged she was both crying and smiling. And the tears were of compassion and the smiles were of hope.

'Ti'Elen gathered her tools and her team about her. They sharpened their chisels, tested their air hammers, cleaned their goggles, and once again, at night, climbed the scaffold and faced the rock.

For long hours, long nights, long weeks, and long months The Building echoed to the sound of figures being released from stone. Once again, a Family appeared, Mother, Father, and Child. But not the Family of Peace and Prosperity depicted years earlier when the work was young. This Family was being torn apart by the strong arm of officialdom while a church burned in the background.

And it came to pass in the early days, not long before the end of New France, that Acadians, choosing not to recognize change, suffered many barbarities. They were burned, herded, deported, and even drowned. And in the long fullness of time Japanese Canadians were torn from their lands and their fishing boats, and entire families were interned. And native children were taken from their parents to be educated in strange cultures and new languages. 'Ti'Elen carved the spirit of the sad visions she had seen, the visions that had misted her eyes with tears of compassion.

But another Family appeared in the stone and this one was making a painful journey, struggling through undergrowth and over obstacles, striving to keep together while striving upward to a safe haven.

And it came to pass that Loyalists fled to Canada from the republic to the south, and before them and after them other peoples came from the far corners of the world – Irish families fleeing famine, Scottish crofters displaced by sheep, English labourers

replaced by machines. It was carved and so it continued. Ukrainians, Poles, Dutch, Hungarians, Italians, Lithuanians, Chinese, Vietnamese, West Indians, Arabs, Indians, Africans – peoples of the world, torn from their roots, came seeking refuge and a new life.

And it was visions of their courage and of the emerging country's outreaching hands and uplifting arms that had caused 'Ti'Elen to smile with hope.

And then 'Ti'Elen carved into the central column of stone that stood almost as tall as herself, and out of the stone a figure emerged. At first it appeared to be a woman, for the robed lines were flowing and the face seemed that of a woman, but there was muscular strength in the arms and in the stance. When the form finally emerged it was that of a man. At his feet lay an iron cage, broken, open, with birds flying free. 'Ti'Elen had carved Freedom.

Freedom is often imagined as a woman, but 'Ti'Elen knew that Freedom must defend itself. She built in muscular strength, and the aggressive will to use it. But the Freedom she created was more than mere liberty. It was a terrible Freedom indeed, for it was the Freedom to Choose.

And it came to pass that the people of 'Ti'Elen's Canada, the Canada that grew from visions and stone and toil, would be able to choose their leaders and their governments, to choose their ideals and their ideologies, to choose Right or to choose Wrong and, having made choices, to enjoy the fruits or to suffer the consequences. It came to pass that Freedom to Choose became the seed, the root, the trunk and the foliage of the many-branched tree that is Canada.

CREATION 85

But 'Ti'Elen stood frightened at what she had done, for in exercising the gift of Freedom to Choose the people could choose to lose that freedom, and without it, all would be lost.

But what was done was done. The fate of Canada was carved in stone. Had she not had visions? Had the sorcerers of the Cone not spoken? For many long years had she and her carvers not created? And was it not good?

*****

The Listeners were silent, as was the ancient teller of tales. Overhead the Moonship had sailed far across the lake, changing masthead stars as it went and laying a path of quicksilver on the deep waters. Boy stood up and walked to the water's edge and shied a flat stone into the night. It skipped disc-like along the Moonpath, stone floating on water, and then, having lost momentum, it vanished forever. He returned and sat down.

Youth put more wood on the fire. Girl went to the tent and returned with a sweater for herself and a blanket for Old One. She spread the blanket across Old One's shoulders and sat down.

"Was 'Ti'Elen's work finished?" she asked.

"Was there a big party?" asked Boy.

"Or at least an unveiling?" said Youth, following Boy's question.

"Ah," said Old One, "you unveil a work of art. You dedicate a monument. Certainly 'Ti'Elen and her carvers had created a monumental work of art, but in doing so they had carved a country. You don't unveil a country. Besides, it had grown at night, over many years. The people beyond The Hill didn't know how it had happened. The Bureaucrats and the Mandarins were too busy trying to manage what had happened.

As for the People's Orators, they wanted everyone to believe that Canada was their creation, as though a country had grown from wind and words."

The Listeners chortled, the fire chuckled, and the loon laughed.

"Just the same," said Youth, "I hope 'Ti'Elen celebrated."

"Indeed she did, and if you listen carefully I will tell you how."

*****

CELEBRATION

PARTY I

"We are talking of celebrations," said Old One, "so let us celebrate with fire."

During the long soft daylight hours of late afternoon, Girl, Youth, and Boy had followed Old One's bidding and had gathered pieces of well dried hardwood. They had collected dried cones and had carried knots of pinewood oozing with gum and had stacked them nearby as Old One had requested.

Now, Old One whispered to Girl and she hurried, wondering, to the cooking tent and returned carrying a clay bowl. Old One took it from her and placed it on stones in the centre of the fire. Old One then cunningly selected sticks from the pile of fuel and slipped them into the already blazing fire, and sparks and flames rushed upward into the night. Each Listener looked at the others through light more living than themselves.

Moonship was still sailing the black vapours of the speckled sky above but was far far from port and the cold light reflected from its pale sails was as nothing compared to the hot light roaring from the fearsome flames.

Old One spoke softly but the fire subdued its own primeval noise to permit the ancient voice to enfold the Listeners.

*****

There was a Mandarin who had been chosen by the People's Orators to rule over the affairs of the Green Chamber. He had watched with wonder and admiration as 'Ti'Elen and her team had carved Canada. Even before 'Ti'Elen had completed Freedom to Choose he had called her to him and had said he hoped she would do something about the great windows that lined the walls of the Green Chamber.

Now it so happened that those windows had been built before anyone knew the full shape of Canada, but it is quite possible the Presence in the Cone had reached out to the architect who designed The Building. It is possible that Essence of Mind had enveloped him just as it had embraced 'Ti'Elen. Many Voice may have whispered to him while the Green Chamber was still merely an idea in his head and pencil marks on flat paper. This we can never know. What we do know is that he built five windows along the West side of the Green Chamber and five windows along the East side and, at the north end over a gallery, two more windows. Twelve windows in all.

And it had come to pass that Canada by now had ten provinces, and the Yukon, and the Northwest Territories. Twelve great jurisdictions, with twelve Gothic window openings waiting to be fulfilled!

The Mandarin of the Green Chamber observed this happy mathematical coincidence, if coincidence it was, and suggested to 'Ti'Elen that she fill each window with symbols to celebrate a province.

And 'Ti'Elen's soul leaped with delight. She could celebrate the parts, and in doing so celebrate the whole. She had carved Canada from solid stone using heavy mallets and hard steel. Now, she would celebrate in pure light and gleaming colour.

Once more she went down the long corridor, the Hall of Honour, that led to the great pine doors. This time she hurried, her mind racing with excitement, eager to embrace Essence of Mind and to confer with Many Voice about her great good fortune.

Inside the Cone all was serene, but no sooner had she entered than she found herself to be no larger than one of the little stone squirrels that scampered forever through the stone foliage in The Building's rotunda.

She was standing shoulder high among the fragile fronds of waving ferns.

The blossoms of wildflowers soared above her on stems like fairy wands. The soft air was scented with the perfume of chlorophyll and the fragrance of humid earth.

She heard at first what sounded like the humming of instruments but gradually the sound that came to her ears was the music of Many Voice reciting in chorus with self.

'Ti'Elen almost laughed. Many Voice, so terrible and so frightening when first heard screaming out of the primal winds of the ages, now sounded like a youngster idling by a trout stream chanting a doggerel ditty to while away the time.

She strained to listen, and this is what she heard:

Two lions and two unicorns,
  stand guard upon the view
and the pretty purple Pitcher
  gathers water from the dew –
while Ostrich is the fundamental frond.

'Ti'Elen did laugh aloud. Many Voice had half crooned half recited the first few lines but the last one had been sung in a deep bass voice that went down a descending scale until the last note seemed to echo in the bottom of a deep well. And the line made no sense.

But did anything else make sense? Lions? Unicorns? A purple Pitcher? An Ostrich? And what was a fundamental frond?

'Ti'Elen had no time to ponder because Many Voice ignored her laughter and crooned on, sometimes using a trilling soprano, sometimes a vibrant alto or a clear tenor, but always on the strange fifth line using a deep bass for a bottomless slide.

On an island in the ocean
    where the warming currents meet
orchids' velvet petals are
    like Slippers for the feet –
while Sensitive's the fundamental frond.

Deep ashore the Trailing Arbutus,
    its leaves forever green,
hides its fragrant pinkish petals
    in the fallen leaves of spring –
while Rusty is the fundamental frond.

Where forests flow to rivers' edge
    and rivers flow with tides
the Purple petalled Violet
    in fragrant coolness hides –
while Maidenhair's the fundamental frond.

In the land where rock and river
    beckon inland from the sea

*the proud Madonna Lily twines
    the formal Fleur de Lys –
while Leather is the fundamental frond.*

*Where fertile field meets rugged rock
    and inland lakes are seas
the Trillium of the wildwood spreads
    a carpet under trees –
while Christmas is the fundamental frond.*

*And on the ancient pastures that
    the bison used to cloak,
the Crocus of the Prairie's
    made of lavender and smoke –
but Berry is the fundamental frond.*

*From lodgepole pine to northern park
    cicadas sing in chorus
and the Lily painted Orange
    paints from tableland to forest –
but Royal is the fundamental frond.*

*Where the grasslands roll forever
    and the mountains frame the west
the Wild Rose of the prairie clings
    to Nature's earthy breast –
but Crested is the fundamental frond.*

*And north beyond the treeline,
    when summer's banished snow,
in bogs and peat and tundra
    the Mountain Avens grow –
but Juniper's the fundamental frond.*

Northwest across the river, past
  the barriers of rock
is the land of Evening Primrose
  and the Fireweed's tall stalk –
but Oak is the fundamental frond.

Then across the mighty mountains,
  the Pacific to the west,
the Dogwood's bracts and berries
  are by setting sun caressed –
and Deer is the fundamental frond.

So sing a song of heroes,
  of battle, trade and questing,
but remember that it's Nature crowns
  the glory, gives the blessing –
Nature crowns the glory, gives the blessing.

All the time, while Many Voice was reciting in a sing song way, little 'Ti'Elen, still no larger than a squirrel in bracken, was seeing ferns and flowers changing around her. From Pitcher Plant, to Lady's Slipper, to Trailing Arbutus, to Purple Violet, the blossoms came and went. Madonna Lilies yielded to Trilliums and Trilliums to Mountain Avens and Avens to Fireweed and Fireweed to Prairie Crocus. Large flowers, small flowers, creeping flowers as high as her knees, tall flowers leaning over her, leaves brushing her cheeks, petals touching her hair, they shimmered from size to size and melted from texture to texture and colour to colour. They grew, withered, died, and grew again even as she watched. Prairie Lily trembled and was gone, Wild Rose taking its place, only to dissolve into mysterious Dogwood whose leaves were like flowers. All this floral pageantry unfolded around her even as Many Voice sang.

No sooner had the simple little ditty come to an end than Many Voice settled in to whistle as carelessly as a small boy going fishing with a willow pole and a bent pin. 'Ti'Elen laughed aloud again and clapped her hands with delight. Many Voice had boomed about "fundamental fronds" in a deep bass voice and what was bass but a foundation line in music? And what was a frond if not the leaf of a fern?

Ferns were the secret!

Wild, native flowers would be the motif but she would make ferns the foundation for her window designs. Ferns would provide weight, balance, and base. The flowers would take care of themselves. Excited, 'Ti'Elen ran from the Cone, not knowing whether she would ever again be permitted to see the interior with the same eyes of wonder.

*****

Old One paused as though to savour the images of the wild garden 'Ti'Elen had just visited.

Girl giggled and spoke up, asking a question that Youth thought, but was too polite to ask. "Did she not look funny scampering down the Hall of Honour no bigger than a squirrel?"

"I'm sure she was full size again," said Old One.

"How?" asked Boy.

"Stories don't need hows," said Old One, grumpily.

"Then why?" asked Boy.

"Why what?"

"Why was she small in the first place?"

"Because we look down on flowers and yet the flowers she would soon create we would always have to look up to. The artist takes a point of view and then gives us her eyes. Just as 'Ti'Elen had imagined history in a long stone mural she now imagined flowers soaring heavenward in glass."

"What *is* glass?" said Boy.

Old One said nothing but added more pine pitch to the fire and sprinkled some powder on it and the flames around the clay pot grew almost too hot to see.

But Girl answered Boy in a superior sort of way. "Silly," she said, "you look through it everywhere. You drink from it every day!"

"Sure," said Boy, unmoved, "but what *is* it?"

"Glass is glass," said Girl, smugly.

"Tell me," said Old One, "what is sand?"

Boy looked puzzled but Youth said, "Sand is rock broken into stone and then ground into fine bits."

"And what does the grinding?"

"Wind and water," said Youth.

"Yes," said Old One and again took the pouch from its deep pocket, and this time a compartment yielded iron-free silica sand that was poured into the clay pot. Another compartment was carefully opened and from it Old One added soda ash to the sand.

"Imagine," said Old One, "that we are ancients by an ancient fire."

"Where?" asked Youth.

"Egypt. Thousands of years ago. Before the temples, before the Sphinx, before the pyramids."

"Before history?"

Old One smiled at Youth, but continued. "We think we are baking clay but have accidentally allowed impurities into the clay – silica sand and soda ash."

The Listeners sat and stared in puzzlement at the clay bowl seated in the white heat of the searing fire. Eventually, Old One took a forked stick and pushed the bowl out of the flames and tipped it over. A clear liquid ran from it and hardened in the cool night air, forming a transparent sheet over the rock of the sloping beach.

"We are looking," said Old One, "at stone that was liquid and now we can see through it!"

"A miracle," breathed Girl.

"No," said Old One, and chuckled, "a mineral. Our ancestors learned to add other minerals such as copper, manganese, iron, even gold. The glass then became coloured – but which colour it became depended on the mixture. Dark blue, light blue, azure blue, turquoise, yellow, and red. But coloured or clear, glass is still liquid stone."

Old One looked at the Listeners and smiled. "Was it not logical that 'Ti'Elen, a sculptor, having carved Canada from solid stone should now celebrate it in liquid stone? Not by making glass, but by using glass?"

Youth looked puzzled. "I don't know. Is it not a different art? How would she learn?"

"Oh," said Old One, "before ever coming to The Hill she had created in stained glass. If you travel southwest from The Hill you will come to The March. There, in open fields, standing beside an old stone rectory, is an even older stone church. Most of its window frames are filled with stained glass pictorials that could be Biblical texts ordered from a glass cutter's catalogue.

"But in the centre of the West wall there is one window that is totally different. Indeed, some members of the congregation dislike it, claiming it is too modern. It is, in fact, delightfully Medieval, not so much in colour as in the design of its figures."

"Is it a picture?" asked Girl.

"Indeed it is. It's called a Resurrection window. At the bottom right three women stand, looking down to the left at an angel seated by an empty grave, and the angel is pointing upward. Above the angel the body of a living Christ is rising upward, one hand reaching as though for heaven, and above Christ, in those heavens, three angels await his arrival and they are looking downward. Their eyes bring your eyes back down to the three women and you are drawn into scanning the scene again.

"It is a window that rotates from Life to Death to the Hereafter and full circle again. Even a non-Christian can understand that window."

Girl was intrigued. "And 'Ti'Elen designed it?"

"Yes, and helped make it."

"Okay," said Youth, "but is it any good?"

"It is there. Judge for yourself. I'm just a Teller of Tales."

"You must have an opinion?"

"It has a primitive beauty that makes me want to tear all the other windows from their moorings and leave the Resurrection floating in space."

Old One poked at the fire and threw on wood that burned with less heat but more light.

"There were other windows, elsewhere, some of which have burned. But now 'Ti'Elen wanted to celebrate Canada in the windows of the Green Chamber. This was a challenge far beyond any windows she had yet designed. How was she to do it?"

"But she already knew," said Girl. "She was going to use ferns and wildflowers."

"That was the subject," said Youth, "not the design." He looked interested.

"True," said Old One, "but even the design she already knew. That was not enough."

"Then what was the problem?" asked Youth. "What did she do?"

"Well," said Old One, "she went to France and sat on the flagstone floor of the Cathedral of Chartres and bathed herself in the beauty of windows that show the life of Christ and the lives of the people, and even dare to symbolize the apocalypse, the very end of the world. Those windows are said to be the most beautiful windows in the world but they were made 800, even 900 years ago!"

Youth seemed disappointed. "Was she hoping for inspiration?"

"No," said Old One, smiling. "She was studying technique. An art lover is interested mainly in *what* is achieved. A creator of art is interested in *how*. 'Ti'Elen was about to create in a cathedral of her own. She had seen the *what* in the Cone. She had drawn sketches. She had made the large cartoons. It was now time for the *how* of it."

The Listeners could tell that Old One was again sliding into the main tale, so they waited, and asked no more questions.

*****

CELEBRATION

PARTY 2

'Ti'Elen had decided that in her cathedral, the Green Chamber, she would celebrate each region of Canada in a garden of glass. Her designs were approved by the Mandarin of the Chamber and by others whose opinion he sought.

But who would actually build the windows? Who would do the cutting, the painting, the staining, the firing, the leading – all the intricate things that had to be done by craftspeople to create the unity of a stained glass window?

It so happened that 'Ti'Elen knew of a fine artist named Russell Goodman, who had built stained glass windows of great beauty. She thought of him, admiringly, as the "Glassmaster." He was now working for a company in The City. The Mandarins and the Bureaucrats conferred and gave a contract to that company to build 'Ti'Elen's windows.

As the work began, 'Ti'Elen watched, and consulted, and conferred, and watched.

Before long she cried in alarm, "This company knows not what it does! These windows are large – over eight metres tall! They are being hung wrong. The weight is enormous. They'll collapse!"

But the Mandarins and the Bureaucrats had made a contract and they had faith in companies and quotes and commerce. And who was 'Ti'Elen but a mere woman who had carved Canada and had had visions of small wildflowers growing tall in a gigantic garden of glass?

'Ti'Elen conferred with the Glassmaster. He, like 'Ti'Elen, had seen the Cathedrals of Europe. He, like 'Ti'Elen, had stood spellbound in stupendous Chartres and had marvelled how the glass made the sun shine inside even on a January day when outside there was no sun. And he, too, was alarmed by the techniques now being used on the Green Chamber windows.

A stained glass window, you see, because it is made up of pieces of glass joined together by strips of lead, is very very heavy. So a window is made in sections not even a metre high but each weighing fifteen to twenty kilograms, and each section is hung from a cross bar of strong steel. The Glassmaster wanted that bar to be "T"-shaped for strength but the Company insisted on a simple round bar.

The Mandarins and the Bureaucrats were unmoved because they believed that companies know what they are doing.

Two windows were built and installed and 'Ti'Elen, who had started out with a celebrating soul, now went into the Green Chamber at every opportunity but with a heavy heart. She sat and stared at the offending windows, pleased with colours and patterns that she knew were the Glassmaster's doing but deeply disturbed by fundamentally flawed techniques that she knew disturbed him, too.

And one day, lo! she saw a sight that pleased her, and her soul immediately felt guilty at her heart's joy. At the top of one window there was daylight between the lead and the stone frame. The window was sagging of its own weight! Once started it would continue in infinitesimal stages as though melting in slow motion and would eventually cascade into the balcony below.

'Ti'Elen made notes and hurried to confront the Bureaucrats.

It is possible her agitation communicated itself to the Presence in the Cone, because in the empty Green Chamber it was as though The Building itself wished to reject the window. Stress gathered on one of the bars. The bar sagged, bent inward, and with a great twang came loose.

It was said that the rod flew like an arrow and embedded itself in the desk of the Queen's First Minister – but tales such as that must be taken as verbal ornaments of a lesser truth.

It was too much for the Glassmaster. He left the Company, saying he would not have his name associated with such work. It was also enough for the Mandarins and the Bureaucrats. They tore up the contract. They listened to 'Ti'Elen. And 'Ti'Elen in turn captured the Glassmaster before he left The City in despair.

'Ti'Elen asked him to be her builder of the windows. He accepted with joy and in turn was accepted by the Bureaucrats.

The Bureaucrats, of course, knowing how eager he was to help create 'Ti'Elen's windows, insisted on paying only a modest amount for each window – in fact, the same amount that had been contracted for in the first place and that had possibly contributed to inadequate installation. Some Bureaucrats, you see, think you can contract art to the lowest bidder because they believe artists work for the love of it. And of course they do – but they also like to eat and to clothe their children. Bureaucrats and Orators prattle about subsidizing art, but usually it is the artist who subsidizes art.

Now it so happened that the Glassmaster had a wife, Nancy, who had studied as an artist and worked with him, and they had three sons who were all still teenagers. Mark, the eldest, was expert at glazing – the fitting of each glass piece into its lead wrapping. Christopher excelled at the delicate art of cutting, and Scott, the youngest, was apprentice to his brothers and his parents. All now came to The City to join the Master and before long they had a studio in an old house not far from The Hill. There they lived and worked.

The boys, being teenagers, could only work summers and weekends, so other workers joined the family as needed, craftspeople like themselves. One was John Kelly and another was George Brunet. The Glassmaster wondered sometimes why glass artisans tended to be wild, even wilder than artists, but these two contributed much laughter, born on an endless stream of jokes. And, whenever her other duties permitted, 'Ti'Elen would join them too.

Every day that 'Ti'Elen entered the old brick house she felt that time had reversed and had carried them all back to a craft home in twelfth-century Europe. It was a Medieval Guild and they were all pouring their souls into their craft and their art to glorify something much greater than themselves. She now thought of Russell as the "Guildmaster."

How does a Teller of Tales describe how this little band turned bits of once-molten sand, and strips of dull lead, into Gothic windows pulsing with light and colour?

Where does one start?

Those windows had begun as pictures in 'Ti'Elen's mind as she stood shoulder high in ferns and saw wildflowers waving above her head.

They became cartoons drawn on huge lengths of paper laid out on a floor. To draw them, 'Ti'Elen would stand upright with her pencil fastened to the end of a long stick, and her arm fully extended. No wrist motion here. The whole motion of her arm flowed along the stick to the pencil and, as she sketched, from the bottom of the cartoon ferns flowed upward, and out of the ferns grew a family of wildflowers in full bloom.

Each cartoon was different. The Dogwood flower of the west coast was on one, growing amidst a fern native to the same area. Next came the Wild Rose of the foothills, with a fern to match, and so on, and on, in infinite variety. Each was different and yet all were the same – wildflowers rooted in a bed of ferns.

Even the youths of the Guild were excited when they saw 'Ti'Elen's designs. Here there was originality of concept, variety within unity, and symmetry shaping chaos.

Here, also, was painstaking detail. 'Ti'Elen sketched the overall flow at arm's length, but she would draw a circle and write "flower," or "leaf," or "frond." Then, to the side, crouched on her knees on the floor, she would carefully sketch detail that made the fern fronds almost curl from the paper, the leaves seeming to spring from the stalks, and the flowers to bloom.

With 'Ti'Elen's designs to guide them it became a matter of pride for the boys, the men, and the woman of the Guild to think only of ways to improve, strengthen, and enhance what was already good, in order to turn a vision into reality. In spite of mythology to the contrary, that is the way much great art is created. One mind may have the vision but many hands and minds create the masterwork.

The first and greatest skill required in making a stained glass window is that of choosing colour and glass. 'Ti'Elen's Guildmaster had a genius for selecting colour. He knew that the French made magnificent blues and that the English made rich golds and rubies, that the Germans knew the secret of superbly warm earth colours, and that in the United States the West Virginian glassmakers produced a rich orange. He only used glass that had been blown, tubed, split and flattened by hand. He loved the imperfections introduced by either accident or cunning. He capitalized on the variations in texture and thickness that characterized hand-made glass.

The Guildmaster took 'Ti'Elen's designs and translated them into cartoons that showed how pieces would fit and where lead lines would be required either to enhance the pattern or physically to support the weight.

These designs were transferred to heavier paper and cut into pieces with scissors that removed a strip equal to the thickness of the lead. This meant that each piece of paper became a precise pattern for a piece of glass.

The Guildmaster would select glass for its colour, its streaks, and its shades, and often 'Ti'Elen would be by his side, choosing this piece or suggesting that. And sometimes they would argue heatedly and other times they would chuckle in agreement, but 'Ti'Elen always yielded to the Master.

A Teller of Tales cannot describe the choosing of colours any more than a bee can describe a sunset. They could have used, if they had wanted, the whole rainbow spectrum of colours but both knew that the miracles of the twelfth century had been achieved with very few colours. The Masters in those days liked reds, blues, whites and yellows. Their secret lay not in the number of colours but in the use of colours. This Master knew that blue worked nicely with white, and that red was happy with yellow, but that even tiny bits of ruby red in lime green made his stomach want to heave.

He knew that, when used correctly and viewed at a distance, red could be made to float right out of blue, as though the window itself had depth. Yellows could be made to float free from brown. Neighbouring colours could be kept separate by the lead and by shading their edges with a brown paint, or they could be allowed to blend, so that where they met the eye would see a third colour.

Such devices are mechanisms built into the work of art to trigger illusions in the brain of the beholder. Such art does not exist in its fullness without the observer.

The streaks and colours of the glass could also be emphasized or altered by using a water paint in which the pigment that gave colour was really fine ground glass. When the painted pieces of glass were baked in a small kiln at high heat the glass pigment melted and fused itself to the main piece. Only fire could ever again separate them.

Sometimes they wanted texture where the glass itself had almost none. The Guildmaster knew many tricks, like sponging the watercolour while it was still wet, or rubbing it gently with his hands and then using a soft brush to remove it. When rubbed and fired it would resemble old glass on which, during the centuries, a spider had spun fine webs that now sparkled in daylight.

And so it went, every piece of glass for a window being chosen, cut, painted, brushed, fired, leaded, mounted into sections, and the sections hung on the bars embedded in the stone frames, until finally a window would be one harmonious whole composed of more than two thousand pieces of glass.

There was a problem with the windows on the West. The powerful afternoon sun beamed full upon them and many years ago the People's Orators had complained of the heat and the light. The old windows had been covered with canvas and heavy curtains had been hung across them.

'Ti'Elen and her Guildmaster knew that if the sun crashed through their windows the Green Chamber would explode with painful light. He suggested painting the outside of the windows with the brown paint and then removing all but a film of it. 'Ti'Elen approved of the idea.

They experimented to find the correct amount of paint and for a while the old brick house was like a laboratory, with artists for technicians. And that is how it came about that the windows on the West wall had a filter fired into them.

With 'Ti'Elen clearing the way through Guards, Bureaucrats, Rules, Regulations, Security, and Red

Tape, the boys and men from the Guild installed the windows themselves. Working from tall ladders and high scaffolds they carefully drilled holes in the stone framework and inserted the "T"-bars. Sections were hung in place and leaded into channels in the stone surrounds.

The boys liked to work outside, high up on the walls, almost under the eaves of the copper-sheathed roof that sheltered the Green Chamber. As they worked they were aware of strange stone creatures that watched them from cornice, corbel, and cranny. Even in areas where few humans would ever go, The Building had eyes.

And so it came to pass that the lovely Green Chamber became even more lovely as glass wildflowers the colour of music cascaded up its tall walls.

Along the eastern wall played the flowers and ferns of Newfoundland, Prince Edward Island, Nova Scotia, New Brunswick, and Quebec. Their colours and tones were muted, to echo the gentle colours of the East and the soft light of the morning sun.

Along the western wall sang the flowers of Ontario, Manitoba, Saskatchewan, Alberta, and British Columbia, the colours growing stronger, deeper, richer, until finally booming with blue and resonating with red jewels, but all modulated by the painted filter.

In the western corner of the northern balcony the many-flowered Fireweed of the Yukon grew in abundance. And snuggled deep into the northern wall of the same gallery the ground-hugging blooms of the Mountain Aven took three small windows that were architecturally framed as four and drew them into one blossoming whole. This was the window of the Northwest Territories and it was 'Ti'Elen's favourite.

"That window," she said to herself, "is the right size for my soul."

"Soul" was the key word. Finally, in the riot of light and colour and blossoms, The Building had acquired a soul.

A keen observer might well have noticed every small animal in every frieze in the building giving a little shudder of sudden animation. 'Ti'Elen had celebrated the Land by celebrating Life, and Life, celebrated, gives Life.

*****

STAINED GLASS WINDOWS

Page 111 – Newfoundland
Page 112 – Manitoba
Page 113 – British Columbia
Page 114 – Northwest Territories

DENIGRATION

Old One fell silent and Youth rose and put more wood on the fire. Youth was being patient with Old One's story but at the back of his head there were questions that, when he had sorted them out, he intended to ask.

Girl sat quietly, her back against a rock, and stared upward at the Milky Way and wondered if it might be a band of sparkling paint coated on the blue of midnight glass waiting for the fire of morning sun. She wondered what skills could construct the constellations and even more she wondered about the Mind that could design them.

As for Boy, his thoughts were deep in his stomach. He got up and went to the cook tent and returned carrying food, and a long fork with many prongs and a wooden handle. Old One rested, Girl marvelled, and Youth pondered, while Boy roasted hot dogs over the fire.

Boy was generous, and fed them each in turn, beginning with Old One, who added so much ketchup, mustard and relish as to make even Boy shudder.

At last Youth, mumbling through the last mouthful of a slightly scorched wiener and a half-burnt bun, asked a question.

"I've een the ill," he said, "and've oored the ing –" He stopped mumbling and chewed instead, then swallowed, and began again.

"I've been to The Hill and I've toured The Building. But when we got to the Green Chamber we had to push our noses against wrought iron and glass. The Orators were not there but we could not go in. Why is that?"

Old One flicked fallen relish into the fire and licked mustard from fingertips before answering.

"They say it is for safety. To prevent vandalism – or worse."

"They 'say'?" Youth was learning to listen to Old One's tones and inflections.

"Remember the guard at the doorway of the Cone? There are always those in authority who manage to limit the vision of the people. The Mandarin of the Chamber who authorized the glass garden was followed by others who tightened the rules of entry. Finally, the people coming to admire the Orators could get in, but those coming to admire art were kept out."

Old One pondered carefully before continuing. "Even the People's Orators had sat spellbound by beauty, their eyes and souls drawn upward by the windows, vision sent winging upward by flowers – but even they had their eyes forced back down."

"How?" said Youth. Boy stopped cooking, and Girl listened with curiosity.

Old One shrugged resignedly. "Philistines hung great bronze chandeliers, each one several metres across, that partially blocked the upward view. They said that even great cathedrals had chandeliers."

"Oh, oh," said Girl. "What did 'Ti'Elen think?"

"She was not happy."

"I'll bet," said Youth, recognizing an understatement when he heard one.

"Particularly," added Old One, "when she knew the Bureaucrats paid twice as much for each chandelier as they had paid the Glassmaster to craft each window."

The astonished look on Girl's face made Old One break into a sudden, cheerful laugh. "Please, child! Understand that when it comes to art those who are not artists think the opportunity to create is reward enough – and that only what is practical has cash value. Chandeliers shed light, so are practical. Windows are valuable only if they let in light and keep out the weather. Why, the philistines ordered storm windows to go outside 'Ti'Elen's flowers – this is, after all, a cold climate – but they paid more for plain glass storms than they did for the stained glass treasures."

Old One was reflective for a moment.

"I tell you this so you have no illusions. You and I know that it is art that tells us what is really important. The mystery is that our leaders don't always know what we know." A sad smile wrinkled the old face. "I suppose I must tell you that next, as though to blind any who saw past the bronze chandeliers, they installed the bright lights of television. And it was ordered that cameras must focus only on Orators and not roam the surrounding beauty of the room."

Old One stopped smiling, and sighed. "Leaders seem to fear the idea of their followers thinking about where they are."

"And *what* they are?" asked Youth, quietly.

"Just so."

Even Boy remained motionless and for long moments nothing more was said. Girl saw that there were tears in Old One's eyes. She sought a distraction.

"I have heard," she said, "that 'Ti'Elen designed windows for the Red Chamber?"

Old One brightened. "The Red Chamber?" Suddenly there was laughter again in the voice. "Ah yes, the 'Other Place' – that's what the Orators call it. The Orators' Chamber is carpeted in green, the Other Place in red. Like Heaven and Hell. Only to be appointed to Hell is close to achieving Heaven, a strange paradox."

The Listeners relaxed. Old One was brightening.

"Yes," nodding at Girl, "quite right. 'Ti'Elen was asked to design new windows for the Other Place, and she did. In the Green Chamber she had celebrated the land with images of wildflowers. In the Red Chamber she wanted to celebrate the land with images of humankind. Does not the land sustain both?" But then, with a slow, sad shake of the head, Old One said, "It was not to be."

"What happened?"

"A great deal, and nothing. The architect who conceived The Building had intended the Other Place to have balconies, like the Green Chamber. They were never built. Instead, the empty expanse of wall beneath the windows had been hung with paintings – fine paintings indeed, but of a ghastly subject. War. Now the Ancient Mandarins of the Other Place said, 'It is time we had the balconies for visitors and admirers, and it is time, too, for beautiful windows.'

"But then 'Ti'Elen listened in amazement as a new architect said, 'No, let us not build balconies, but let us paint balconies!'"

"Paint balconies!" exclaimed both Youth and Girl, and even Boy looked amazed.

"Yes. 'Let us paint balconies,' he said, 'but so cunningly they will fool the eye.' Of course he had seen palaces in Europe with painted pillars so real you would swear they held up the ceilings, and painted doorways so real you would break your nose trying to walk through them."

Girl giggled as she pictured onlookers falling off painted balconies.

"Now," said Old One, smiling, "it is quite possible the Mandarins of the Other Place were lacking in imagination. It might have been most suitable to have cunningly painted balconies to remind us that nothing is ever exactly what it appears to be." Old One chuckled. "In either Chamber there is always the possibility of appearances having no substance! That architect may not have been as mad as the Ancient Ones of the Red Chamber supposed. However, their minds were sealed up."

Old One sighed. "And then along came a gaggle of artists from another City and demanded a country-wide competition to see who should design the windows."

Girl was shocked. "Windows 'Ti'Elen had already designed!"

"Indeed. But a few of the Ancient Mandarins of the Other Place knew, just as 'Ti'Elen had always known, that an artist creating a masterwork within The Building must never create it for himself or herself. Art in The Building was not there to glorify the artist but to illuminate The Building. And The Building was a growing thing that had begun life generations before and would continue to evolve for generations to come. The artists of each generation working within The

Building had to let their own work grow out of that of their predecessors."

"How is that possible?" said Youth. "What about artistic freedom?"

"What about it?"

"How can an artist be shackled by the past?"

"Ah, indeed," said Old One, casting a warm look of appreciation upon Youth. "What a puzzle. Most good artists would agree with you."

"'Ti'Elen was good!" Girl was almost defiant.

"There is an excellence that transcends good," said Old One, patiently. "In The Building it was not the Past that shackled. Nor was it the Mandarins, or even the Bureaucrats. The artists who were humble enough to understand The Building shackled themselves.

"Consider. When she carved history, 'Ti'Elen used Romanesque, a rough style that was not Gothic. But her windows she made flow with naturalness – Gothic windows are usually stiffly stylized and full of symbols. Yet all was in harmony with the Gothic building. It's not conformity that's required, but harmony."

"What's Gothic?" asked Boy, surprising Old One, who had suspected Boy was interested only in the architecture of hot dogs.

"Yes," said Girl. "You hear of Gothic horror stories. What's the connection?"

"Well," said Old One, "originally Gothic meant savage, crude, rough. And in a way a Gothic building is that. But it also has high pointed arches over doorways and windows. Inside, its vaulted ceilings seem to soar. Outside it has steep roofs with high gables, and above the roofs rise tall towers and graceful spires. Often the walls have so many tall window openings that they have to be supported by flying buttresses, great arched legs that brace the walls so they don't spread outward."

"Like the buttresses on the Cone?" asked Girl.

"Indeed."

"Then Gothic is easy enough to spot?"

"Ah, but it's much more than that. It has variety. It gains uniformity by defying uniformity. It is not a shape found in Nature but it is decorated with natural forms. A Gothic building is wreathed with intricately carved foliage, and animals, and people. There will be portraits of humble folk alongside caricatures of the high and mighty, both cheek by jowl with gargoyle monsters. A Gothic building is utterly rigid but thanks to the carvings the very stones that give it rigidity are seething with activity. For many modern artists repetition is a crime and excess is a sin but a Gothic building thrives on excess and its ornamentation accumulates like the swelling notes of a symphonic finale. No matter how the notes thunder the soul yearns for more. Any artist who understands Gothic can work within Gothic."

Youth looked thoughtful. "And is that why 'Ti'Elen had no grand unveilings? No showings – no –" he searched for words.

"No ego trips," said Girl.

"Yes," said Youth.

"Ah," said Old One, very pleased. "Those who work successfully in Gothic subordinate themselves to Gothic. The Building is, was, and always will be more important than the individual artist. You know the story of Faust?"

Girl looked puzzled. "Didn't he give his soul to the Devil?"

"Yes. In return for favours."

"Are you suggesting," said Youth, "that The Building was 'Ti'Elen's Devil? What favour did it promise?"

"A secure career as a creative artist doing what no artist had ever done before – carving a country."

Youth was intrigued but disbelieving. "And she gave her soul in return?"

"No. She gave the one thing many artists can't do without. Fame. Gothic denies fame."

"Didn't she long to break free?"

"Of course. But she was exercising freedom, was she not?"

Youth looked puzzled. It was Girl who exclaimed, "Freedom to Choose!"

Old One smiled. "'Ti'Elen had learned with much inner struggle and soul-searching that she must subdue her artist's ego. She had to give of herself, wholeheartedly, not asking for fame in return.

"As 'Ti'Elen she never so much as carved her initials into a crevice in a piece of stone. True, the Glassmaster painted her name and his side by side on stained glass and fired it in place forever, but the lettering was so small and so high up the window that it could only impress a wandering bat. Besides, there were those television lights searing the eyes of the beholder."

Old One chuckled. "If an installer of a window carved his name in the outer lead that was then pushed into channels in the stone frames, what was the harm? That was not ego. It was like giving even more of oneself to The Building. But then, remember, along came those stained-glass artists from another City demanding a great competition to choose 'the best' person to design new windows for the Red Chamber.

"They went further than that. These were artists for whom only the abstract had merit. 'Ti'Elen's genius lay in being able to reach the hearts and minds of ordinary folk like us. Her recognizable flowers offended the snobbery of the abstractionists. They denounced the windows in the Green Chamber, saying they were a plague upon The Building – proving in one word that they didn't understand The Building."

"I wonder – ," Boy mused, "what would Many Voice have said to them?"

"I don't think," said Old One, "they could have heard Many Voice. Or would have listened if they had."

"Poor 'Ti'Elen," said Girl.

Old One almost exploded. "Poor! Poor? Because of a little criticism? If it had swayed her from her path one inch, then say 'Poor 'Ti'Elen'!

"Anyway, I am sure Many Voice was not consulted by the protesting artists. And the Mandarins of the Other Place were too annoyed to carry on. In fact, what with architects talking of painted balconies and invading artists prattling of plagues, the Ancient Mandarins shut 'Ti'Elen's window designs away in a large drawer."

Girl was sad. "Where are they now?"

"In a large drawer. But don't mourn for 'Ti'Elen. By this time she was trying to figure out a purpose for the Ancient Mandarins, as well as for the People's Orators. She had carved Canada but that wasn't enough. A country's no use if it can't be managed. How on earth do the people govern a country like

Canada? How do they organize themselves? What rules do they follow? How do they make those rules? How do they enforce them? In short – how do they unite themselves to avoid social chaos?"

Old One peered at the Listeners and again asked, "How do you do these things?"

"I know," said Girl proudly. "With laws!"

"Yes, but what Law rules the rulers?"

There was a long silence before Youth offered a suggestion. "A constitution?"

"Ah, indeed. So 'Ti'Elen, having carved Canada, now had to carve a Constitution."

"Where?" asked Girl.

"Have you forgotten? There were twelve great blocks of blank stone flanking the windows in the Green Chamber."

Girl did remember. "Yes! She'd been shown them when she first arrived. She was told they were for her!"

"Well, that was where she was going to carve a constitution – with mallet and chisel – in stone."

Youth laughed.

Old One looked annoyed. "That strikes you as funny?"

"Not at all. I'll bet the abstractionists weren't calling for a competition on that!"

Old One smiled. "I can tell you this – 'Ti'Elen went again to The Cone, and this time in fear and trembling. Not for fear of what she might see but for fear that she might find nothing. Anyone setting out to carve a constitution needs inspiration. And who better to provide it than Essence of Mind speaking through Many Voice?"

UNIFICATION

STEP 1

The interior of the Cone had changed. Before, there had been nothing identifiable as an "interior." Now, however, 'Ti'Elen had the distinct impression she was walking on a fine floor that was intricately patterned with inlaid strips of contrasting hardwoods. There were still no limits and perhaps there never would be, but here and there she came upon desks and tables and narrow iron staircases that seemed to beckon upward to enticing regions above.

She wanted to explore but her progress was impeded by a throng that appeared to have gathered for a meeting. Never before had she seen such a strange group.  There was a lion, wearing a crown, who walked with a peculiar hopping motion because he insisted on going on three legs while holding a leaf-sceptre in his right front paw.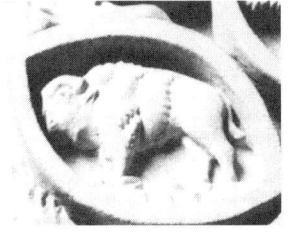

There was another lion who walked around on his hind legs, trying to look tall and imposing.

There was a big bull buffalo who watched 'Ti'Elen sideways, while a tall moose with spreading antlers stared

directly at her in a most disconcerting way.

There was a bear wearing a ruffled collar that made him look very princely and he strolled around looking straight ahead, talking to anybody who wanted to talk nose to nose and ignoring everyone else.

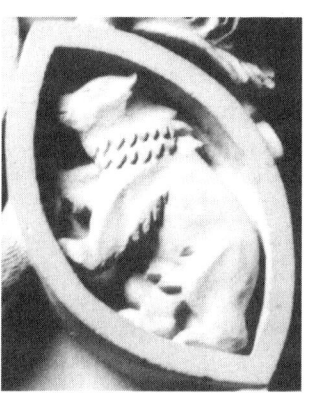

There was a husky dog with a big ruff around his neck, a majestic owl with wise eyes and a vicious beak, and a unicorn with an elegant horn sprouting from its forehead.

They were all talking to each other at once, in a friendly sort of way but with a lot of passion and sometimes in very loud voices that made 'Ti'Elen rather nervous.

While trying to make her way around the owl without ruffling his feathers she almost fell into a pond. Nearby, a beaver and a seal were playing on the beach. On the far shore two little twin mountains rose upward, as trim and neat as the teeth on a saw. The pond was salt water and the wind of the discussions raging beside it caused waves to wash back and forth. In the centre of the pond was a little island, with trees. At one end of the pond a small galleon, under full sail, battled through the heavy seas while at the other end Old Father Sun had hung his crown in the air and was taking a bath in the salt water without raising so much as a wisp of steam.

Beside the pond grew lilies and wheat, and there was wheat bundled into sheaves. And all the living things were talking to each other! The island trees whispered to the wheat, the wheat to the lilies, the lilies to the bear. The bear growled to the lion, the lion to the moose. Even the little boat seemed to be talking to Old Father Sun.

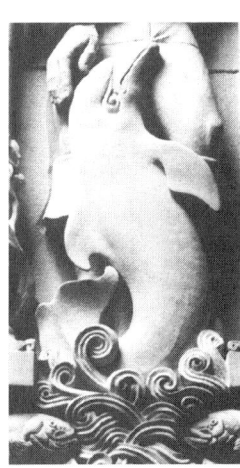

Around and around in the pond swam a dolphin. Above the dolphin, making wide looping circles in the air, flew a fierce looking falcon. The dolphin and the falcon seemed to be talking to everybody simultaneously in two languages. All the conversations were broadcast as an echo by the twin peaks. The babble of voices was the strangest thing 'Ti'Elen had ever heard and she couldn't make sense of any of it.

In the centre of the Cone she saw the white pillar that she had seen once before, and again, as before, she took refuge near it, only to find that it was not a pillar at all, although it was indeed made of white marble. It was the tall figure of a Queen and it rose high above everyone's head, even that of the moose. The marble Queen, however, was not talking. She held her head slightly tilted upward, as though determined to remain aloof, high above the herd.

Suddenly 'Ti'Elen laughed, for she realized what was going on. Essence of Mind had peeked inside her head and had seen that her sculptor's mind was full of visual symbols of Canada and so, like a prankster, Essence of Mind was letting those symbols loose all about her. The babble of voices was merely Many Voice in a playful mood.

'Ti'Elen was now quite comfortable with Many Voice and had ceased being either frightened or intimidated. In fact, if the truth were known, she had become quite fond of Many Voice.

'Ti'Elen moved away from her sanctuary at the feet of the marble Queen and, trying not to be distracted just because it was a moose talking, or a three-legged lion, or a lily, she tried to unravel the conversations. They went something like this –

"I can get along quite nicely by myself." – "No you can't, you're not big enough to defend yourself without us." – "That's true." – "Not one of us is rich, but together we are rich." – "That's for sure." – "Together we can build things we all need." – "A throne would be nice." – "What we need is a Federation!" – "No, no! A Confederation!" – "What's the difference?" – "Who cares as long as it works?" – "Just so long as I can do what I want." – "No! Let's give most power to the federation and just some power to each of us. That way we have to stick together." – "That's not a bad idea." – "I'd like something in ermine with a sceptre." – "Power? What do you mean?" – "How about gathering taxes?" – "Power to spend taxes, that's the ticket!" – "Power to educate, that's real power!" – "A crown would be nice." – "Who will make the laws?" – "Who will say if we've broken the laws!" – "What language should we speak?" – "Whatever the language, we need the right to say whatever we want." – "Every person must have equal power." – "That's right. One person, one vote!" – "No, no! If a majority gang up they can always rule all the others." – "Then we need votes for regions, not people?" – "We need both!" – "And who protects the people from the lawmakers?" – "*I DO!*"

There was a sudden silence and everyone looked startled. "Who said that?" asked the unicorn.

"I did, and I do," came the reply.

Everyone looked up in surprise and above them the marble statue of the queen nodded ever so slightly, as though to say, "Yes, you heard me."

Now that she appeared to have everybody's attention she spoke again, in high clear tones that flowed as gently as water in a woodland stream.

"I will be the symbol of the people," she said. "Let their power flow through me in trust. Let no one govern without my approval. Let no law be passed without my signature. Let my presence remind all who are chosen to govern that they are chosen by the people, that they rule for the people, and that they can be removed by the people. And to show that I am a symbol of the people's power while being personally powerless, I recommend you choose someone to represent me, who will act for me, so that I who symbolize the people's power can never be corrupted by touching that power."

"A democratic monarchy!" said the bear.

"Of course," said the three-legged lion plaintively, "that's what I've been trying to tell you."

The Queen, who held a scroll in her right hand and a wreath in the other, extended her right arm at full length and pointed the scroll at 'Ti'Elen, but not imperiously. She was more like a mother addressing her eldest daughter.

"Attend to it," she said.

And suddenly 'Ti'Elen was alone.

The parquet floor of the Cone stretched away on all sides to real walls that were lined with shelves made from richly carved wood. The shelves rose three floors high and were reached by the little iron stairways. At the second and third levels, narrow walkways with floors made of wrought iron and glass gave access to the shelves. On the shelves, cover to cover and row above row, were books, books, books, thousands of books. Above the last shelf of books, far above 'Ti'Elen's head, the walls rose still higher, pierced by slender windows, then continued to rise above that, sloping inward, supported by an intricately groined ceiling, finally meeting in a lovely golden dome suspended high above the marble figure of the queen.

Just at that moment, as 'Ti'Elen surveyed the remarkable room, the door from the Hall of Honour opened and a young guide brought in visiting tourists. They crowded forward to stand near the middle of the room.

"This room," said the guide, "is the library. It was built more than a hundred years ago. It was the only thing to survive the great fire."

'Ti'Elen smiled to herself. A lot more had survived the fire than a mere room, lovely though it was. Whatever else might go on in the rest of The Building she knew the Cone would continue to shelter Essence of Mind, who would always speak through Many Voice to anyone who cared to listen.

She smiled at the statue of the long-dead queen and quietly thanked the long-dead sculptor who had created it, then, leaving the library, went to attend to her appointed task. In her head were images of animals, and people, and heraldic symbols, all of which were part of a sculptor's vocabulary. That vocabulary now had to be used to express some very complex ideas.

'Ti'Elen knew that the nature of the vocabulary, when combined with the nature of the great slabs of stone that protruded from the walls of the Green Chamber, meant trouble for carvers. That stone was full of fossils of ancient sea creatures and was very difficult to carve. Not only was the material extremely hard but fossil chunks were likely to pop out, leaving unwanted blemishes in the work. She wanted to take those stones right out of the wall and replace them with a limestone more friendly to the chisel.

There was only one problem.

If she had the men remove the stone slabs, would the Green Chamber collapse?

You see, no one knew precisely how the walls were structured. What was below the surface? How deep were the slabs? Were they mere ornamental blocks or were they part of the structure?

The Building had been designed by an architect, so once there had been plans, but the Orators had shown no interest in mere blueprints and the Bureaucrats had ignored an opportunity to preserve them. A glance at those plans would have told all, but now the cold stone told nothing. It was the kind of challenge 'Ti'Elen enjoyed.

She and the men took power drills into the Chamber and drilled test holes to find the true dimensions of the huge stones and to find how they were seated and secured. Imagine her delight to learn that the architect had foreseen her problem many years before and had ensured that those stones could be removed and replaced. Because that was so, it would not be necessary to carve them in place.

'Ti'Elen visited quarries in search of stone that would both welcome the chisel and blend with the fossil-bearing stone of the Green Chamber walls. One does not select stone slabs weighing many tonnes from a supply on the shelves of a craft shop. Eventually she found what she wanted. It was Oolitic limestone and it, too, had been formed millions of years ago when the pressure of water in a great inland ocean had crushed tiny marine creatures into stone. But this stone was kindlier to the chisel than the Tyndal stone that lined most of the interior of The Building. 'Ti'Elen had it shipped to a new workshop in an old warehouse not far from The Hill. The carvers were no longer working in the sub-basement of The Building.

But some of the carvers had gone.

Iti Busolo was gone, as was Anton Nielson, Fernand Rossignol, Roland Rossignol, Wilfrid Filion, and Joe Joanisse. But Joe Joanisse was here in spirit in his two sons, Maurice and Marcel, who had been apprentice carvers during the final years of the Creation of Canada. And a bearded young sculptor from South Africa joined the team. His name was Christopher Fairbrother.

'Ti'Elen had Christopher take his tools to the south gallery of the Green Chamber, where blank stone corbels protruded from the wall. They were to be embellished as an exercise in the sheer pleasure of carving. 'Ti'Elen suggested that one be turned into a dragon, for no reason other than that she was a romantic at heart and liked old legends. Christopher carved a fine dragon.

Christopher too was fond of legends and knew of one about a mischievous character called Jack o' Green who liked to play tricks and cause problems. So he carved an apple tree with a man picking apples. But the man was falling because his ladder was moving. The ladder was being moved by Jack o' Green. And that is how Jack o' Green, the Mischief Maker, came to the Green Chamber and he has been there ever since.

But 'Ti'Elen was busily sketching designs to represent a Constitution that, if at all successful, would prevent too much mischief from entering the affairs of Canada.

While she sketched, great blocks of Oolitic stone were being brought into the large workshop in the old warehouse. Here the men used huge power saws to cut the stone into thick slabs the same size as those on the walls of the Green Chamber. Some were for carvings that would be as much as four feet wide by six feet high and two feet thick. Such a carving would weigh many many tonnes and so right from the start each was composed of three blocks of stone. The three blocks were set one above the other on massive wooden easels so that they could be carved as a whole. Eventually, a fourth, smaller, decorative stone would be inserted at the foot of each sculpture.

'Ti'Elen knew that it would take at least a year to make an entire carving so she decided to relinquish control over much of the detail. The overall concept would be hers but the men would be allowed much more freedom of expression than she had given them in the Green Chamber foyer.

And so it began.

A Teller of Tales cannot recreate the sight of that workshop with its gargantuan easels, tall stepladders, writhing air hoses, and dust-laden air. Nor can a Teller of Tales make you hear the scream of the saws, the rumble of exhaust fans, the rattle chatter and buzz of power-driven chisels, or the whine of drills and the tap tap tap of mallets on steel. Oh for the skill to help you imagine the figures of 'Ti'Elen and her carvers, clothed in overalls, eyes behind goggles, air filters over mouth and nose, ears muffled in defenders, as they set to work. Those who think of artists as soft creatures afraid of hard work, and who imagine a Constitution as being no more than a creation of wind and words, could do well to think of 'Ti'Elen and her little team of carvers as they set their hearts and muscles to a task that would take the next ten years of their lives.

*****

Old One paused, eyes staring off into the darkness as though trying to recapture long-gone images.

"Do go on," said Girl to Old One, then, suddenly worried, "Are you tired?"

"Not at all, but it's getting late."

"The moon is still listening," said Youth. And indeed it was, although more distant than ever.

"Ten more years," said Girl, trying to imagine such a spread of time. "How long had 'Ti'Elen already been on The Hill?"

"Twenty years."

"Gee," said Boy.

"Still unknown?" Youth seemed to have a problem with 'Ti'Elen's lack of fame.

"Not totally," said Old One, "but in general, yes."

"Then I don't understand," said Youth. "I still think artists crave acknowledgement. How could she – well – you know – keep going?"

Old One sighed. "And how do I explain? All the time 'Ti'Elen was trying to perfect her art she was also trying to perfect herself. In subtle ways." Old One turned to Boy. "Tell me. Right now. What would you most like to have?"

"This minute?"

"Yes."

"A chocolate bar."

Girl laughed but Old One shushed her. "A good answer. With caramel, and nuts?"

Boy's mouth almost watered.

"I have one in the tent," said Old One.

Boy's eyes sparkled.

"But tell me," said Old One, "do you really need it? Really?"

Boy was tempted but settled for honesty. "No."

"Then can you do without it for now?"

Boy looked dejected but nodded. "Yes."

Old One did not relent. "Tell me, when you said you could do without, did you not hurt just a little inside?"

Boy looked surprised, but nodded.

"Of course. In fact, that was a tiny little death. Soon you'll get over it. Soon you'll feel better for having disciplined yourself into denial. You'll feel so good it will be like a little rebirth. And it could only happen because of the death. That was something 'Ti'Elen had been learning."

"She liked chocolate bars, too?" said Boy, his eyes wide with pleasure.

Youth, Girl, and Old One all laughed.

"What you're saying," said Youth, "is that she wanted recognition but turned her back on it –"

"Yes, and died a little in doing so."

"– but was reborn stronger?"

"Indeed. And she would have loved to have carved every inch of Canada by herself but time and human strength would not permit. With every piece of stone she relinquished to a carver she died a little death, only to be reborn stronger by taking pleasure in the accomplishments of the men.

"She had been tempted to fight the Bureaucrats over the chandeliers and the TV lights, but she didn't. In fact, throughout all the long years she seldom fought with the Bureaucrats. She learned to be patient, to deny her ego and her anger, to die a little – and wait for rebirth. She wanted a husband and a child but pushed that want aside – a big death. She pushed artistic temperament to one side – a little death. She told herself she did not require public adulation – a bigger death. Jack o' Green encouraged an architect to attempt to have her removed from her job, so she had to deal with anger and frustration – and die again. The more difficult the deaths the bigger and the more satisfying were the rebirths. It's an old principle. Do you understand?"

"No," said Boy, but Girl nodded and Youth looked thoughtful.

"You see," said Old One, "little deaths and little rebirths, the rebirths accumulating year after year so that the person grows, and, if the person is an artist, then the art should grow, too – that's what she believed and so that was the way she tried to live."

Old One pulled the blanket closer and Boy leaped up and brought more wood.

"You can go to bed if you wish," said Old One to Boy as the youngster stoked the fire.

"And miss a good fire?"

"True," said Old One. "Missing a fire is like missing life. Both burn for such a brief time. And don't forget that her accident had made her very much aware of time. The voice by the roadside had told her she still had work to do, and it was still not done."

"I'm not tired, either," said Girl, hearing the storyteller's rhythms slipping into gear.

"The night's still young," said Youth.

Old One smiled at the courteous lie and continued.

*****

UNIFICATION

STEP 2

'Ti'Elen had carved a Canada that had grown from a glaciated barrens inhabited only by a few isolated cavemen to a verdant country sparsely occupied by sophisticated native peoples to a land supposedly discovered by European explorers and finally populated by people from many parts of the world. But the population was still not large. And the land mass was enormous. Visitors could cross the Atlantic Ocean from Europe to eastern Canada and still be closer to home than eastern Canadians would be who had merely journeyed to the western shores of their own country.

How was such a land to be governed? Could it be governed? It was obvious that if everyone was going to work together for the good of all they would have to follow the same set of general rules – a Constitution. Once again, 'Ti'Elen with her pencil had to make sense of everything she had seen, heard, learned and thought so that the chisels could then shape the Ideas that would in turn shape a Constitution.

I will not tell you precisely how it was done, or in detail whose chisel did what or in what order. I leave it to you to imagine the skills required to be carving fine figures one moment and the next to be hoisting tonnes of stone onto scaffolds built against the walls of the Green Chamber. I leave it to you to imagine 'Ti'Elen and her team removing the big blank blocks of stone and replacing them with carved gargantuan slabs as neatly as putting a piece into a puzzle.

I leave it to you to imagine the years rolling by until one day it was all completed.

And what did the Orators have, and the Mandarins, and the Bureaucrats, and we the People?

Well, for one thing, 'Ti'Elen had remembered how big the country was and how varied were its geography and its people. She also remembered that it had grown in almost natural stages, like branches on a tree, with each branch being unique but receiving nourishment through the trunk. So she carved the idea of a Confederation that would be like a tree, in which a single trunk could carry individual branches bearing fruit that would be the Yukon, the Northwest Territories, and the Provinces of Newfoundland, Prince Edward Island, Nova Scotia, New Brunswick, Quebec, Ontario, Manitoba, Saskatchewan, Alberta, and British Columbia.

Because she used her sculptor's vocabulary this unusual tree carried the moose and the bear and the boat and other symbols, each identifying a fruit on the tree of Confederation.

But she remembered that the strange creatures in the Cone had all agreed the Confederation should be a Monarchy, with a king or a queen holding power in trust for the people, so beneath the Confederation tree 'Ti'Elen placed the lion walking on three legs and holding the royal sceptre.

This was the "Confederation Stone," and it came to pass that Canada became a confederation of governments each with many powers of its own but all receiving vitality from a common central government,

that in turn received its power from a Queen, who received her power by consent of the people.

'Ti'Elen remembered the words of the marble queen and so carved another stone in which she imagined a representative of the Queen, seated on an almost royal throne, reading a royal statement. This person was a Governor General, not royal but symbolizing royalty. Powerless but a channel for power.

Again she used her sculptor's vocabulary and put a roof over the Governor General's head to show that the royal authority extended both indoors and outdoors. Above him she put both a beaver and a seal to show that the same influence extended over land and water. Again she used the tree to say that all political authority flows from a common origin – the people.

She showed the Governor General standing beneath the tree and acknowledging and encouraging the achievements of others.

At the base 'Ti'Elen thought she would carve something strictly for decoration, but she carved a unicorn, the symbol of purity, and in doing so carved a lasting reminder to all Governors General that they symbolize a power that must be above corruption.

This was the "Governor General Stone," and it was one of several stones that 'Ti'Elen carved almost entirely by herself.

Now it's all very well to say that power passes from the people, through the Crown in trust, to the Governor General and on to the men and women who actually exercise the power. But how were we to select these men and women?

The answer lay in the Vote, and in the theory that since all people are equal each person should have one vote.

Again the chisels rattled away and a strange design appeared. At first glance one could have said that 'Ti'Elen's mind must have been off in the land of the ancient Aztecs, for there was something ornately rigid and mathematical about the design. But on close observation one could see rivers of water all converging on a central point. Around the perimeter were the faces of people representing men and women of many racial origins, and each face was beside a river path and all paths led to a central point that was a square, or box, with a ballot in it. On the ballot was an x indicating an individual vote. The rigid pattern of this carving suggested that ideally every adult must have equal access to the ballot box and that the vote must

be used very carefully and precisely. It was not something to be abused or thrown about at random.

The mathematical harmony of the main carving was reflected at the base by a human harmony. There, in a stone that fairly flowed, four children sang in chorus.

Christopher carved all the faces. Into the mouths of the singing children he seemed to carve the words, "O-Ca-na-da."

This was the "Franchise Stone." And it came to pass that the vote, or franchise, is not just a symbol but is the single most important tool for the use of people who wish to govern themselves. Without it, how would they control the use of power?

But 'Ti'Elen realized that the vote is merely a device by which power is passed from many to a few. The few who are elected to wield power must have a place in which to gather and confer. They must have rules by which they debate so that even when they disagree fists never replace words.

So 'Ti'Elen sketched again, and this time she imagined an ideal People's Orator. She saw this person with three faces, one looking backward into the Past, one looking ahead into the Future, and one looking directly at the Present. She imagined the Orators being elected by the people but knew that when they met together they would need a chairman, so she drew in a Speaker. And someone to keep notes – a Clerk. And someone to enforce order – a Sergeant-at-Arms. They would also need a certain number of Orators present to make any meeting legal – a Quorum. She included a figure to represent the leader of a government, the Queen's First Minister, or Prime Minister. More than that, she included a figure on the other side as the leader of the Opposition. This recognized the fact there are always people who dislike what a government is doing, and that they must have someone to express their opinions.

'Ti'Elen thought this would be a good stone to give to Maurice Joanisse, who was no longer an apprentice, but a skilled carver. She stepped aside and, of course, died a little.

Maurice also carved a base stone and at the bottom he showed people holding an election to send Orators to a Parliament, and the Parliament was in The Building on The Hill.

And it came to pass that the Green Chamber became known as the House of Commons, because the theory was that the men and women elected to go there represented the common folk of Canada. And those representatives became known not just as Members of Parliament, but as Members of the House of Commons.

The "House of Commons Stone" is a reminder to the elected Members that if they cease to represent the common folk then they have no right to be there.

But what if the Members were to forget whom they represented? Or what if those from the heavily populated areas of the country were to gang up on the others who were less numerous? Tyrants are not always single dictators or ruthless emperors. The majority can also rule tyrannically. 'Ti'Elen knew that thanks to the vote a government could be overthrown without bloodshed, but it would take time, and much damage could be done in the meantime.

'Ti'Elen pondered over this paradox, the paradox that democracy has the potential to be tyrannical.

Again she put pencil to paper and eventually charcoal to stone, and symbols for another chamber emerged. There was an owl, and a woman holding juggling balls, and a man with a fulcrum, all of which symbolized the need for wisdom, mental dexterity, and balance. In the base stone there appeared a long table with people seated about it engaged in deep discussion. This symbolized the need for a great deal of study and thought, because the men and women in this chamber would have the power to change laws proposed by the House of Commons, or even to prevent them being made.

Thus it came to pass that the Senate was established to counter-balance the House of Commons even though the Senators were not elected. The "Other Place," The Red Chamber, housed this Senate. It too had to have a quorum. Figures symbolizing this were carved by Maurice.

While this "Senate Stone" was being carved Jack o' Green whispered playfully into Maurice's ear. Maurice, amused, created a senator in golfing clothes eager to be elsewhere, another so old and tired he was leaning on a cane, and yet another hurrying as though having forgotten a meeting.

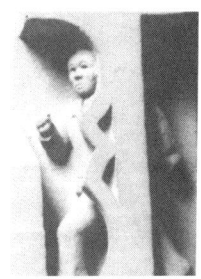

And it came to pass that the Senate was not a perfect institution. Of course, in 'Ti'Elen's philosophy, imperfection meant there was room for growth.

Creating a Parliament with both a House of Commons and a Senate was only a beginning. 'Ti'Elen had to think about the powers of Parliament, because that is what a constitution must define. She knew that throughout history there was one power beyond all others that separated the rulers from the ruled. It was a power that the British Parliament long ago had taken away from its king, making it very clear that the people were in charge. It was the power to collect and spend money. It was the power to tax.

And so again 'Ti'Elen's charcoal flew over the stone blocks in the workshop and again the compressors hissed and the air chisels chattered. This stone she gave to Marcel, Maurice's brother, and under his hand symbols appeared suggesting some of the areas in which Parliament should spend money. Health,

defence, resources, native peoples, immigrants, the nation – many such things grew from the chisel blade. Above those symbols was a family, large and central, – a Mother, Father, and Child. It was, after all, for their benefit that taxation had been invented.

The "Taxation Stone" was designed to remind the Members that although taxation is government's most important power, service should be government's most important objective.

At the bottom Marcel carved a Blue Whale, an endangered animal. It was a way of saying that the environment that sustains us all must be of major concern to Parliamentarians spending the people's money. 'Ti'Elen always liked to show Marcel's Blue Whale to visitors.

'Ti'Elen gave Christopher a tall, broad stone, told him the theme, made some suggestions, and left him to it. A nurse appeared, with a patient. A man in a helmet wearing tanks on his back was soon cutting into steel with a blowtorch. A sculptor swung mallet on chisel. Money changed hands. A girl played beautifully on a violin. Children clustered around a teacher. A professor wearing a hat with a flat top read from a

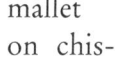
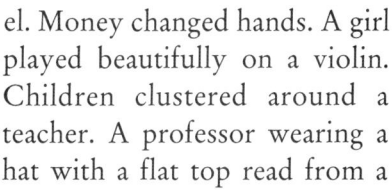

scroll. Amidst all this activity were the Provinces and the Territories and the Yukon. Behind it all, supporting it all, was the ever present Tree of Confederation.

Christopher had carved Education into the Constitution, itself a subject so complex that it would affect all of government.

In the base stone there appeared a fearful scene. A magnificent owl, wings spread as it landed, was caught by Christopher's chisel in the very act of seizing a mouse. One set of talons ruthlessly killed the victim and the other set clawed forward over the very edge of the stone. This exquisitely savage scene was at the base of the "Education Stone" to remind the Members that education is neither an ornament nor a luxury but vital for a nation's survival.

While the stone chips flew in the workshop and the scaffold hoists strained and creaked against the walls in the Green Chamber, 'Ti'Elen's mind continued to ponder the scope of a Constitution that must define the powers of governments. She thought of children, the victims of divorce, and she thought of automobile accidents like her own. She thought of fire and of arguments over insurance claims. She pictured a farmer objecting that a new dam would deprive his cattle of water while a businessman argued that without the dam there could be no electricity.

She pictured all such arguments being taken before a judge in a court to find a fair and just solution.

To handle such problems Parliament would have to write Civil Laws and set up Civil Courts. And it was done.

This "Civil Law Stone" was carved by 'Ti'Elen, Maurice, and Christopher.

While pondering Civil Law 'Ti'Elen had been angry, thinking of her own accident years ago and of subsequent problems with insurance companies, and so she designed imposing pillars that represented wealthy companies as well as judicial courts. But by the time she was carving the base stone she had simmered down and was even laughing. In that stone she created a harmless collision between an old automobile with wooden wheels and an ancient bicycle called a penny-farthing. A dog, running to watch, thought the scene amusing.

And it came to pass that Civil Law deals even with things that might seem trivial, for it deals with daily life.

But daily life, and what other life is there, can also involve murder, robbery, rape, vandalism, and many other kinds of violence. Canada would need laws to guide the police and the courts in preventing or punishing such crimes while at the same time protecting citizens from being falsely convicted of crimes.

'Ti'Elen designed a "Criminal Law Stone" showing a judge, and two people charged with crimes. One man, innocent, was being set free. The other, guilty, was in chains. Criminal Law, too, became part of the Constitution.

Christopher carved this stone but Jack o' Green must have sneaked in with pencil, charcoal, mallet and chisel, for the judge held a gavel, something not used

in Canadian courts. And in the base stone, beneath a Latin word meaning "Truth," a Mountie was chasing a fugitive who was big and bold while the Mountie was small and insignificant, a relationship that 'Ti'Elen felt to be the opposite of truth.

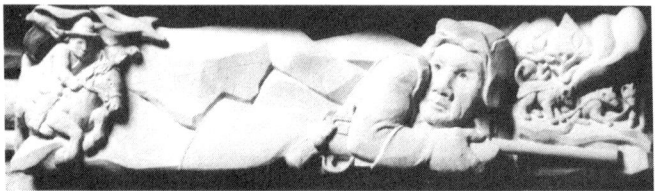

'Ti'Elen's artistic instincts wanted to replace the base stone but she refrained – and died a little.

Jack o' Green must also have been nudging one of the Orators, for a Member rose on a point of order, and complained that at the top of the stone there was an arched roof with rioters struggling up one side of it carrying a sign saying "We want," while from the other side police and firemen were pouring water on them to beat them back. "This kind of thing does not go on in Canada," said the Member, and he demanded that the entire sculpture be removed.

A journalist scurried away to do research and wrote a scathing article proving the Member to be sadly wrong. The Stone remained firmly embedded in the walls of the Green Chamber just as Criminal Law is embedded in the Constitution.

The many ideas behind a Constitution were gradually taking shape. Every day, even while the Members debated in the Green Chamber and the Senators conferred in the Red Chamber, a few blocks away in the dusty noisy workshop 'Ti'Elen and her assistants were busily creating.

It occurred to 'Ti'Elen that nothing they had done to this point made any sense at all without another major ingredient. Even her experiences in the Cone would have been meaningless without it. It was something that separated humankind from all the other animals. Without it, Canada could not function, nor could the Law, or Parliament itself. It was a thing so vital to human life that all governments would have to preserve it, encourage it, and regulate it.

And so 'Ti'Elen designed a stone with symbols that looked like balloons, boats, helicopters, electrical cables, a telephone, and radio telescopes. Between the telescopes she placed a symbol representing the ability to analyze the composition of the stars. A big letter A was joined to a large z by a railway train. Large in the bottom half, two humans talked to each other face to face, and framed in an oval above their heads an ancient printing press had a place of honour. In the smaller base stone, mail was being delivered to a rural box. This was the "Communications Stone."

It was a complex stone, cluttered, and lacking in fluidity and grace. But it came to pass that Communications would become one of the most complex subjects for governments to deal with. Communications would cut across all levels of society, and if it should be exploited or corrupted, Freedom itself would be threatened.

The Communications Stone was carved by 'Ti'Elen, Maurice, and Christopher. "Too many hands," thought 'Ti'Elen. But she kept her peace – and again, died a little.

However, while pondering the complexity of Communications and thinking back to the final panel of History in the Commons lobby, 'Ti'Elen realized that the fundamental means of communication was human speech, and that the people coming to Canada from faraway countries must be able to talk to each other.

She remembered that even though Canada was home to many races, the citizens of two European empires had helped found it. At that time those two empires encircled the globe and also embraced many races. But each empire had one major language that its people used. The language of one people was English, and that of the other was French.

'Ti'Elen almost laughed, for she remembered the Dolphin swimming in the pool in the Cone, babbling cheerfully away in two languages to the Falcon wheeling overhead. "My goodness," she thought, "those two languages that are already in worldwide use – they should be the languages that assist communication in Parliament and throughout the trunk of the Tree of Confederation."

She went cheerfully again to her drawing board and then to a stone. She reached far back into her memory for medieval symbols. A person appeared dressed in heavy work clothes. It was Everyman – you and me and all others. The figure was standing on water and above its head there was a roof, which was the sculptor's way of saying that everything this stone symbolized applied indoors as well as out of doors. Two pillars supported the roof and up one pillar twined the Lily of France and around the other spiralled the Rose of England. Across the top of the stone curled flowering Lupins, a plant that grows everywhere in Canada, even in the Arctic. The Lupin meant universality.

Swimming in the sea at the base were fish, symbolizing that Everyman had had to cross water to arrive in Canada. But now, out of that same sea there rose a sleek Dolphin and it stood on its tail to greet Everyman, who placed his right hand in a friendly fashion on the Dolphin's head. At the same time a great Falcon settled onto Everyman's upraised left hand, and in this way the sculptor let the Dolphin of

France and the Falcon of England say that Canada should have two official languages – English and French. This was the stone of two tongues, the "Bilingual Stone," lovely in concept and beautiful in execution.

'Ti'Elen, whose mother tongue was English, carved Everyman with the Dolphin and the Falcon, while Maurice, whose mother tongue was French, carved the water, the fish, and the flowers.

And then 'Ti'Elen, wondering what to carve into the ornamental base stone, was inspired to give it to Maurice to do with as he pleased. Maurice carved an Inuit hunter in a kayak, caught at the very moment of hurling a spear at a walrus. Arctic water rippled beneath and jagged ice rose behind.

It was ornamentation, but it said that no matter their race and no matter their region, the people of Canada would communicate with the Federal government in either one of two languages, for the Constitution is not History, it is Law.

'Ti'Elen looked at the artistry of the little base stone and was very happy. She knew at that moment that Maurice was a sculptor. Many little deaths flowered into rebirth.

There was a twelfth stone still mute, still staring blindly from the wall of the Green Chamber. 'Ti'Elen had an opportunity to carve one more constitutional principle into the law of the land. What should it be?

She thought of her strange experiences in the Cone and of the enormous variety of thoughts, ideas, opinions and concepts that had flowed into her through Many Voice. It occurred to her that if Many Voice had been forbidden to speak to her, she, 'Ti'Elen, would never have been able to carve Canada.

And so this stone, too, she eventually took to herself. Again a roof appeared near the top but this one implied thought, learning, and the freedom to teach. It, too, was supported on either side by pillars that at first glance appeared to be wreathed with Gothic foliage until a close look showed the foliage to be composed of people of many races and occupations, all ribboned together with spiralling conversations. Standing tall and strong between the pillars and beneath the roof there grew the figure of a woman wearing a flowing gown. She was Freedom of Speech.

If 'Ti'Elen had been creating a constitutional arch, Freedom of Speech would have been the keystone at the peak, holding it up and locking it together.

But 'Ti'Elen did not stop with the figure of the woman. Across the top of the "Freedom of Speech Stone," almost hidden from everyone's view, she carved a serpent. And it came to pass that the Serpent of Evil always lurks near Freedom of Speech and that Canadians would always be torn between suppressing the one and protecting the other.

*****

EVOLUTION

Old One stopped talking and for a few moments no one said anything. Out in the lake a hunting fish surfaced, making a small rippling splash. Deep in the woods a horned owl hooted and all was again silent. Boy lay beside the fire, curled up, full of marshmallows and hot dogs. He was fast asleep. On the horizon Moonship was just drifting into harbour, about to pull the moonpath behind it. Suddenly, close to shore, Youth saw a small v-shaped ripple of water moving across the dwindling trail.

He gestured toward sleeping Boy, then rose quietly and very gently tossed a pebble, aiming it to land beside the moving v. Instantly there was a splash like the sound of a gun going off as the beaver, for that is what it was, splashed its tail in alarm and dove for safety. Boy woke up, startled. Girl laughed, and Old One smiled.

Youth lifted Boy to his feet and guided him off to the sleeping tent.

Girl put wood on the fire and spoke to Old One.

"I could sit up all night."

"Fine," said Old One. "It is good to sit all night by a lake."

"And make up stories?"

"If you wish. As for me, I never make up stories."

Girl sat for a long time, not knowing what to think.

Youth returned and stood on the beach watching Moonship's sails being lowered onto a distant deck. At last he turned and rejoined Old One and Girl by the fire.

"I would think," he said, "that 'Ti'Elen must have been very content?"

"Contentment," said Old One, "is not a natural condition for an artist."

"But she had carved Canada!" exclaimed Girl and there was a smile in her voice. "And a Constitution! And she'd celebrated in glass! You yourself said so, and you never make up stories –" she let the tease hang in air like a lure and hoped Old One would rise to it.

"What has been made up? She carved history into the long stones of the foyer and did it not come to pass? Is Canada not all around us? She celebrated using the ancient art of stained glass and is there not beauty all around us? She carved ideas into the stones in the Green Chamber and do those ideas not shape our lives?"

"But what came first?" Girl had an orderly mind.

Old One answered carefully. "It has been truly said that in the beginning was the Word – a poetic way of saying in the beginning was the Idea. It is quite possible to have all kinds of happenings, events, actions, without having a country. A gaggle of people occupying a certain land mass and organizing themselves so they can fill their bellies and bury their excrement does not make a country. It's Ideas that make a country. Ideas that are expressed in Mythology. And it's artists like 'Ti'Elen who create the Mythology."

"But what," said Youth, suddenly intrigued, "– what if the artists won't – or can't?"

"Why then," answered Old One, pulling the blanket close as though to shut out the very thought, "there is no country."

There was a long silence broken eventually by Girl. "I am glad that 'Ti'Elen came to The Hill."

"Yes," said Old One, "and give thanks for those who came before her and those who will come after."

"Will there be others?" Girl was surprised.

"Others! Even while she was on The Hill there were carvers in the Arctic, by the Pacific, in the mountains, in lake country, all carving stones at home to embellish The Building. They were commissioned by the Mandarins but 'Ti'Elen would travel to visit them when they needed encouragement or guidance. She'd be on hand to help install their finished work. These carvers had names like Sioui, Akulukjuk, Jacobs, Anghik, and Akiterk. They all carved their own Mythology. By carving it into The Building they helped carve it into the Idea of Canada."

Youth understood. "Like poets and writers," he said.

"Of course. An artist is an artist. They may use different vocabularies – words, colours, music, shapes, textures – but a shape in stone can be a symbol for an idea just the way a word is a symbol for an idea. But in The Building on The Hill something else happens. Artists come and go and each one adds to the Gothic whole. Individual artists depict their individual visions but what gradually emerges is a Collective Idea of the country. When that happens and when the people understand it, the country has happened."

Youth suddenly saw that Old One had walked into a trap. "So then," he said, "'Ti'Elen did not carve Canada. They all did."

Old One smiled, quietly pleased that Youth was alert. "A little exaggeration is a Story Teller's privilege. Besides, hers was a major vision, and it continues to grow."

"What do you mean?" said Girl.

"Well, it could be that 'Ti'Elen paid close attention to what the Aboriginal carvers had to say. Anyway, she didn't stop when the Constitution was carved. Oh my no. It's quite possible that she felt a little guilty about making everything begin with people – remember the Cavemen in the lobby?"

"But that was after Many Voice had been howling at her," said Girl, defensively.

"Certainly. But now she had found her own voice. And her own eyes. She looked around the Green Chamber and saw great blank triangles of stone over the arches that framed the visitors' balconies. Those stones begged to be embellished.

"While 'Ti'Elen was thinking about those stones Jack o' Green came along and whispered in one Mandarin's ears that those same stones should carry symbols of cherub-like faces blowing the four winds. It was a motif once popular in elegant ballrooms. 'Ti'Elen was appalled by the idea and fortunately so were other Mandarins. They told her to carry on and to do what she thought best."

"Which was?" Youth had been thinking of going to bed but now he changed his mind.

Old One glanced at the stars the way one looks at a clock, then nodded and continued.

*****

'Ti'Elen thought of her first terrifying visit to the Cone and how Essence of Mind had whirled around her and Many Voice had howled and jeered until, in desperation, she had stoutly said that a country begins with people. And Many Voice had said, "Very well, begin with people."

But now 'Ti'Elen wondered if Many Voice had merely been helpful, trying to force her hand to begin somewhere. Few artists can face a blank stone or even a blank piece of paper without feeling terror at the thought of where to begin. And so 'Ti'Elen had begun with Cavemen.

Now it could be that since then she had been affected by all the flowers she had studied for the windows. Or it could be that she had pondered over Marcel's Blue Whale. It might even be that she had noticed that so many of her own sculptor's symbols were figures of creatures like moose, bear, beavers, owls, and falcons. In The Building these were symbols carved in stone, but in the vast territories now ruled from The Hill they were real living creatures who once had been quite independent of humans.

Throughout all the many years that 'Ti'Elen had moved throughout The Building she had also noticed that the People's Orators in the Green Chamber had usually acted as though they were the lords of creation. As though they ordered the rising of the sun and the going down of the moon. As though the great river that flowed at the back of The Hill flowed past at their request.

So 'Ti'Elen created designs that would remind the Orators that life did not begin with humankind. That it began to evolve millions of years before humans ever set foot on the planet, let alone found an ice bridge from Asia to North America.

Once she began to think about evolution she found it difficult to stop, because who can conceive of an ending? Surely evolution begins with physical things but then becomes a matter of the mind and has to do with the creation of abstract thoughts that become universes in themselves. And so 'Ti'Elen continued to sketch, finally even designing a sculptor's impression of the Quantum Theory, which itself is an idea so abstract that a Teller of Tales would not attempt to put it into words, let alone stone!

But who would do the carving? I remind you that 'Ti'Elen was no longer young.

I also remind you that the wall of History in the lobby of the Green Chamber had a unity of concept to it that was very pleasing. It also had a unity of style, for in general the hand that carved the major figures had been the hand of 'Ti'Elen.

And unity of concept and tightly disciplined execution made the stained glass windows in the Green Chamber a joy to behold.

But 'Ti'Elen herself had reservations about the Constitution Stones. She had relinquished some of the design and much of the carving to others, feeling it was their right to participate fully. A few of the stones were ugly. All were interesting. Some were elegant. Amidst the wealth of detail were small gems – Marcel's whale, Christopher's owl and his singing children, Maurice's Inuit hunter.

'Ti'Elen thought of the work that lay ahead in order to create a series of stones depicting "The Evolution of Life" and was determined that unity of concept should again be blessed by unity of touch. How she longed to pick up mallet and chisel and go to work.

How she longed even to participate in the actual sculpting! But no, her time on The Hill would soon be over.

She decided that this final series of stones would be carved by one hand, that of Maurice, who had begun

as an apprentice, become a carver, and was now admired by 'Ti'Elen as a sculptor. The evolving artist would carve evolving life.

And so Maurice Joanisse set to work with 'Ti'Elen's designs and under 'Ti'Elen's guidance.

As the noisy dusty days rolled by, a small sea organism appeared above one arch, an intricately

coiled mollusc above another. An ancient armadillo shared space with a star sponge and a magnificent sabre-toothed tiger snarled down from its own roundel. By yet another arch there marched a triceratops carefully placing its feet in proper triceratopal order.

As each stone was finished it, too, was hoisted into place in the Green Chamber.

'Ti'Elen hoped that eventually the People's Orators would be surrounded by enough history and symbols and ideas and beauty that on the days when they felt truly despondent they could lift their eyes above the television lights and the chandeliers and be inspired to participate in the ongoing processes of evolution.

'Ti'Elen had been hired to create some Gothic embellishments but 'Ti'Elen had carved Canada and had inserted it into a universe. Of course, that's what a Gothic building is – a universe in stone.

*****

Old One fell silent and, after a moment, stood up. "You two can stay up, but I'm for sleep."

"Do you need anything?" asked Youth.

"Sleep," came the answer and Old One moved off into the shadows. For just a moment Youth thought the Teller of Stories vanished before reaching the tent, but he knew it had to be a trick of the firelight.

Girl and Youth sat side by side throughout the remainder of that night listening to the loons and watching shadow islands afloat on the gentle water beneath the drowsy star eyes of the sky.

They thought of the Firebird coming with coals in its talons to create the world, and of the structure that died in fire only to be reborn as The Building. They thought of 'Ti'Elen coming to The Hill and how she suffered many little deaths of self-denial only to be rewarded with rebirth and many accomplishments.

Youth and Girl thought of these things in different ways but both dreamed the same dream, which was that they, too, would participate in the ongoing creation of Canada.

INDEX OF CARVINGS AND WINDOWS

 29 Primitive Man, Woman & Child
 31 Inuit Hunters
 32 Indian Hunters
 33 Vikings
 36 John Cabot
 37 Cabot's Ship
 38 Jacques Cartier
 40 Journey to the Interior
 41 Champlain and Friend
 48 "veined hands grasping"
 49 Capture of Quebec
 49 Deer and Vines of Life
 50 Kings Exchange Treaties
 51 New France to British Colony
 53 Settlers: Man, Woman and Child
 54 Invasion Repelled
 58 Christianity Arrives
 59 Organized Religion
 61 Tree of Knowledge
 62 Teacher
 63 Students
 66 Red River Settlers
 67 Map Maker David Thompson
 68 Through the Rockies
 69 Railroad to the Pacific
 70 Exchanging Trade Goods
 71 Banker
 75 Strangers in Their Own Land
 76 The North Pole
 77 Law and Order
 79 Exploiting Nature
 82 Violent Separations
 83 Seeking Refuge
 85 Freedom
 111 Newfoundland

112 Manitoba
113 British Columbia
114 Northwest Territories
125 Three Legged Lion
125 Upright Lion
125 Buffalo
126 Moose
126 Bear
126 Unicorn
127 Dolphin
141 Confederation Stone
142 Governor General Stone
144 Franchise Stone
145 Franchise detail
146 House of Commons Stone
147 House detail
148 Senate Stone
149 Senate Stone details
150 Taxation Stone
151 Taxation detail (whale)
151 Education details
152 Education detail
152 Education Stone
153 Education detail (owl)
153 Civil Law Stone
154 Civil Law detail
155 Criminal Law Stone
156 Criminal Law detail
157 Communications Stone
160 Bilingual Stone
161 Bilingual detail
163 Freedom of Speech Stone
171 Sea Organism
171 Mollusc
172 Armadillo
172 Sabre Toothed Tiger
173 Triceratops

MUNROE SCOTT is among the most dedicated of freelance writers and artists in this country. Having begun as a staff writer with Crawley Films in the early 1950s, he is perhaps best-known as the writer and director of the CBC-TV series *The Tenth Decade; First Person Singular* (The Pearson Memoirs); and *One Canadian* (The Diefenbaker Memoirs). In addition to writing the *Sound & Light Show* for Parliament Hill, 1984-93, Munroe Scott is a biographer of Dr Robert McClure, and an award-winning playwright and columnist.

After studying film at York University, IAN SCOTT became a freelance lighting technician working on feature film locations as far afield as Budapest, the Great Wall of China, and the jungles of Borneo. He then moved behind the camera as an independent Director of Photography. His foray as a still photographer into the Centre Block of Canada's Parliament Buildings is only one phase of what is proving to be a fascinating and wide-ranging career.

PENUMBRA PRESS